# Tissue Engineering Made Easy

# Tissue Engineering Made Easy

Edited by
## Farhana Akter
University of Cambridge, Cambridge, UK

ELSEVIER

AMSTERDAM • BOSTON • HEIDELBERG • LONDON
NEW YORK • OXFORD • PARIS • SAN DIEGO
SAN FRANCISCO • SINGAPORE • SYDNEY • TOKYO

Academic Press is an imprint of Elsevier

Academic Press is an imprint of Elsevier
125 London Wall, London EC2Y 5AS, UK
525 B Street, Suite 1800, San Diego, CA 92101-4495, USA
50 Hampshire Street, 5th Floor, Cambridge, MA 02139, USA
The Boulevard, Langford Lane, Kidlington, Oxford OX5 1GB, UK

**Library of Congress Cataloging-in-Publication Data**
A catalog record for this book is available from the Library of Congress

**British Library Cataloguing-in-Publication Data**
A catalogue record for this book is available from the British Library

ISBN: 978-0-12-805361-4

For Information on all Academic Press publications
visit our website at http://www.elsevier.com/

  Working together
to grow libraries in
developing countries

www.elsevier.com • www.bookaid.org

*Publisher:* Mica Haley
*Acquisition Editor:* Mica Haley
*Editorial Project Manager:* Tracy I. Tufaga
*Production Project Manager:* Anusha Sambamoorthy
*Cover Designer:* Maria Ines Cruz

Typeset by MPS Limited, Chennai, India

# CONTENTS

# LIST OF CONTRIBUTORS

**F. Akter**
University of Cambridge, Cambridge, United Kingdom

**L. Berhan-Tewolde**
St Mary's Hospital, London, United Kingdom

**N. Bulstrode**
Great Ormond Street Hospital, London, United Kingdom

**A. De Mel**
University College London, London, United Kingdom

**H. Hamid**
Health Education East of England Deanery, Cambridge, United Kingdom

**J. Ibanez**
Guy's and St Thomas' Hospital, London, United Kingdom

**M. Kotter**
University of Cambridge, Cambridge, United Kingdom

# ACKNOWLEDGMENTS

I would like to thank my coauthors—Mr. Bulstrode, Dr. De Mel, Dr. Hamid, Mr. Ibanez, Mr. Kotter, and Dr. Tewolde-Berhan.

Thank you Mr. Amin Ali for providing illustrations for this book.

I would also like to thank the students I have taught, who motivated me to write this book.

I would like to thank the staff at Elsevier, who helped me through the editing and production stages.

Finally, I would like to thank my loving family. I dedicate this book to my parents—thank you for all the sacrifices.

# EDITOR BIOGRAPHY

Dr. Farhana Akter is currently a doctoral student at the University of Cambridge and a surgical resident. She has a Master's in Reconstructive Surgery, and is the recipient of numerous awards for her work during medical school and postgraduate training. Her avid interest in regenerative medicine and surgery has kept her on the front lines of medical research, and her passion for education and teaching has led her to present and lecture both nationally and internationally. She has organized multiple international medical conferences, and while still an intern published her first medical book, *OSCE PASSCARDS for Medical Students.*

Tissue engineering is a fascinating and complicated subject at the cutting edge of medical and surgical innovation. It is the hope of the author that *Tissue Engineering Made Easy* will bring a basic understanding of its core principles to those new to the topic, and promote further reading and research in the field.

**Farhana Akter, MBBS, BSc, MSc, MRCS**

# CHAPTER 1

# What is Tissue Engineering?

**F. Akter**
University of Cambridge, Cambridge, United Kingdom

## 1.1 INTRODUCTION

The term "tissue engineering" was officially coined at a National Science Foundation workshop in 1988. It was created to represent a new scientific field focused on the regeneration of tissues from cells with the support of biomaterials, scaffolds, and growth factors (Heineken and Skalak, 1991).

Tissues or organs can be damaged in various ways, such as by trauma, congenital diseases, or cancer. Treatment options include surgical repair, artificial prostheses, transplantation, and drug therapy. However, full restoration of damaged tissues can be difficult, and the resulting tissues are not always functionally or esthetically satisfactory. The damage to tissues may be irreversible, and can lead to lifelong problems for the patient. In such cases, organ transplantation can be lifesaving; however, this is greatly limited by the lack of donor tissue. Surgeons therefore face a number of challenges in reconstructing damaged tissues and organs.

Tissue engineering enables the regeneration of a patient's own tissues, and thus provides the potential for reducing the need for donor organ transplants. It also reduces the problems faced with traditional donor organ transplantation, such as poor biocompatibility and biofunctionality, and immune rejection. However, despite extensive animal research, human studies are limited. Although tissues such as skin grafts, cartilage, bladders, and a trachea have been implanted in patients, the procedures are still experimental and costly. A focus on low-cost production strategies is thus critical for the successful mass production of effective tissue-engineered products. Solid organs with more complex histological structures—such as the heart, lung, and liver—have been successfully recreated in the lab, and although they are not currently ready for implantation into humans, the tissues can

Tissue Engineering Made Easy. DOI: http://dx.doi.org/10.1016/B978-0-12-805361-4.00001-1

be useful in drug development and can reduce the number of animals used for research (Griffith and Naughton, 2002).

In the following chapters we discuss the tissue engineering applications available for different systems of the body, and their relevance to clinical practice and surgical treatment.

## What is Tissue Engineering?

1. The use of a combination of cells, engineering materials, and suitable biochemical factors to improve or replace biological functions.
2. An interdisciplinary field of research that applies both the principles of engineering and the processes and phenomena of the life sciences toward the development of biological substitutes that restore, maintain, or improve tissue function (Langer and Vacanti, 1993).

## What is Regenerative Medicine?

Regenerative medicine refers to both cell therapy and tissue engineering. Cell therapy utilizes new cells to replace damaged cells within a tissue to restore its integrity and function. Tissue engineering encompasses three approaches: the use of bioactive molecules such as growth factors that encourage tissue induction; the use of cells that respond to various signals; and the seeding of cells into three-dimensional matrices to create tissue-like constructs to replace the lost parts of tissues or organs (Howard et al., 2008).

## REFERENCES

Griffith, L.G., Naughton, G., 2002. Tissue engineering—current challenges and expanding opportunities. Science 295 (5557), 1009—1014.

Heineken, F.G., Skalak, R., 1991. Tissue engineering: a brief overview. J. Biomech. Eng. 113 (2), 111—112.

Howard, D., Buttery, L.D., Shakesheff, K.M., Roberts, S.J., 2008. Tissue engineering: strategies, stem cells and scaffolds. J. Anat. 213, 66—72.

Langer, R., Vacanti, J.P., 1993. Tissue engineering. Science 260, 920—926.

# CHAPTER 2

# Principles of Tissue Engineering

**F. Akter**
University of Cambridge, Cambridge, United Kingdom

## 2.1 INTRODUCTION

Tissue engineering (TE) provides opportunities to create functional constructs for tissue repair and the study of stem cell behavior, and also provides models for studying various diseases. In order to produce an engineered tissue, a three-dimensional environment in the form of a porous scaffold is required. The construct also requires appropriate cells and growth factors, forming the TE "triad" (Fig. 2.1). The cell synthesizes new tissue, while the scaffold provides the appropriate environment for cells to proliferate and function. Growth factors facilitate and promote cells to regenerate new tissue. It is important to tailor the components of the TE triad for specific tissue applications. Each component is individually important, and understanding their interactions is key for successful TE (Grayson et al., 2009; Jakab et al., 2010).

Tissue engineering has allowed the successful creation of isolated constructs, and also hollow organs such as those found in the cardiovascular (see chapter: Cardiovascular Tissue Engineering) and respiratory systems (see chapter: Lung Tissue Engineering) and the gastrointestinal tract. The gastrointestinal tract can be engineered to repair damage caused by diseases such as stomach cancer and inflammatory bowel disease, and to replace sphincter tissue. Numerous studies have used collagen scaffolds seeded with intestinal smooth muscle to create intestinal tissue. However, it has been difficult to create a tissue that mimics the natural contractile function of the smooth muscle cells in vivo (Hendow et al, 2016). Replacing sphincter tissue can reduce the significant morbidity associated with fecal and urinary incontinence in patients. Animal studies have shown that scaffolds seeded with mesenchymal stem cells can lead to improved leak pressure in a rat urinary incontinence model (Shi et al, 2014). However, fully functional tissue-engineered rectal

Tissue Engineering Made Easy. DOI: http://dx.doi.org/10.1016/B978-0-12-805361-4.00002-3

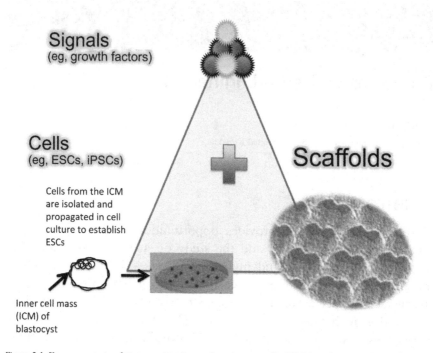

*Figure 2.1 Key components of tissue engineering: embryonic stem cells (ESCs) and induced pluripotent stem cells (iPSCs).*

sphincters have yet to be created. Bladder cancer affects millions of people worldwide and the current standard of bladder reconstruction using the intestine leads to a number of complications. There is thus a huge need for an alternative method of bladder tissue replacement. Bladder engineering requires biomaterials that can regenerate the urinary bladder, allow direct cell—cell interaction, and reduce scar formation. The scaffold needs to be seeded with cells (eg, mesenchymal stem cells) that can differentiate into a number of cell types found in the urinary tract, such as urothelial smooth muscle and neuronal cells. More research in large animal models with long-term follow-up is required before this method can be translated into a clinical setting (Adamowicz et al., 2013).

## 2.2 SCAFFOLD

Scaffolds serve as artificial extracellular matrices (ECMs) used to support formation of the tissue-engineered tissue. Similar to natural ECMs, scaffolds are used to assist with proliferation, differentiation, and synthesis of cells (Chan and Leong, 2008). To fulfill the functions of a scaffold in TE, the scaffold should meet a number of requirements (Table 2.1).

| Table 2.1 Ideal Properties of Scaffolds Used in Tissue Engineering (O'Brien, 2011) | |
|---|---|
| **Properties** | |
| Biocompatibility | • Scaffolds must be biocompatible to allow cell adherence onto surfaces. <br> • Scaffolds must elicit minimal immune reaction. |
| Biodegradability | • Scaffolds must be biodegradable to allow cells to produce their own extracellular matrix (ECM). <br> • The degradation by-product must be nontoxic. |
| Ideal mechanical properties | • Scaffolds should have mechanical properties consistent with the anatomical site into which they are to be implanted. <br> • Scaffolds must be strong enough to allow surgical handling during implantation. |
| Ideal architecture | • Scaffolds should be porous to ensure adequate diffusion of nutrients and cells within the construct and to the ECM. |

## 2.2.1 Biomaterials

An important component of the early advances in TE was the parallel development of artificial biomaterials. According to the European Society for Biomaterials, a biomaterial is a "material intended to interface with biological systems to evaluate, treat, augment or replace any tissue, organ or function of the body." Scaffolds are typically made of biomaterials. These include naturally derived materials, synthetic polymers, and acellular tissue matrices. Naturally derived materials and acellular tissue matrices obviate the problem of immune rejection. However, synthetic polymers can be reproduced on a large scale with controlled properties (Olson et al., 2011).

### Synthetic polymers

Synthetic polymers represent the largest group of biodegradable polymers. They are biologically inert, and their properties can be tailored for specific applications. The most common synthetic polymers used in TE are poly-L-lactic acid (PLLA), polyglycolic acid (PGA), and poly-DL-lactic-co-glycolic acid (PLGA) (Nair and Laurencin, 2007). PLLA has good tensile strength, but it degrades slowly. PGA and its copolymers (such as PLGA) have good mechanical properties, but degrade too quickly when used as a scaffold; thus their tensile strength reduces by half within 2 weeks (Ikada, 2006).

### Natural materials

Natural materials can be classified as proteins (silk, collagen, gelatin, fibrinogen, elastin, keratin, actin, and myosin), polysaccharides (cellulose, chitin, and glycosaminoglycans), or polynucleotides (Nair and Laurencin, 2007).

**What are the advantages of using natural materials?**

Natural materials are biologically active and promote cell adhesion and growth. They are biodegradable and allow host cells to produce their own ECMs and replace the degraded scaffold (Nair and Laurencin, 2007).

**What are the disadvantages of using natural materials?**

Natural materials can induce a strong immunogenic response, and there is the possibility of disease transmission (Nair and Laurencin, 2007). Most naturally occurring materials yield products with low mechanical strength in comparison to synthetic polymers (Ikada, 2006).

**Acellular scaffolds**

Acellular scaffolds are ECMs that have been prepared by removing cellular antigen components from tissues to reduce immunogenicity. Acellular ECMs have been used in scaffold-free TE of many tissues, including lungs, heart valves, vessels, and skin (Chan and Leong, 2008).

**Scaffold-free tissue engineering**

Scaffold-based constructs face a number of specific challenges. These include potential immunogenicity, uncontrollable degradation rate, and toxicity of degradation products that can affect the long-term behavior of the engineered tissue construct. Scaffold-free TE refers to any platform that consists of just cells and signals. Scaffold-free approaches can be engineered by exploiting cell–cell adhesion and the ability of cultured cells to grow their own ECMs, thereby helping to reduce and mediate inflammatory responses (Jakab et al., 2010). A variety of techniques have been developed to engineer tissues without any scaffold, with notable success in tissues such as musculoskeletal cartilage. However, scaffold-free TE has yet to provide a method where constructs can be produced with high output (Norotte et al., 2009).

**Scaffolding approaches and fabrication techniques**

Over the last two decades, four major scaffolding approaches for TE have evolved (Table 2.2). Polymer scaffolds must possess many key characteristics to be useful as materials for TE. These characteristics are determined by the scaffold fabrication technique (Table 2.3).

## Table 2.2 Scaffolding Approaches (Chan and Leong, 2008)

**Scaffolding Approaches**

| | |
|---|---|
| (1) Seeding cells in pre-made porous scaffolds | Porous scaffolds are fabricated before cell seeding. |
| (2) Decellularized extracellular matrix (ECM) | This removes cellular antigens (reduces immunogenicity) from the tissues, but preserves the ECM components. |
| (3) Confluent cells with self-secreted ECM | Cells can be cultured on a polymer until confluence. The cells secrete their own ECM, and membrane sheets form. The cell sheets can then be detached and implanted. |
| (4) Cell encapsulation | Cells are entrapped, usually in hydrogels, before being injected. |

## Table 2.3 Scaffold Fabrication Techniques

**Fabrication Techniques**

| | |
|---|---|
| Solvent casting/particulate leaching (SCPL) | • Polymer dissolved in organic solvent.<br>• Particles that act as porogens (eg, salt) are added.<br>• Solvent evaporates, leaving a polymer matrix with embedded particles.<br>• Composite immersed in water.<br>• Salt leaches out, producing porous structure.<br>(Mikos et al., 1994) |
| Gas foaming | • Polymers are exposed to gas foaming agents (eg, carbon dioxide) at high pressures until the polymers are saturated with $CO_2$ and a foam entity results.<br>• The pressure is reduced, resulting in the release of $CO_2$ from the polymer.<br>• Nucleation and growth of pores in the polymer.<br>(Sachlos and Czernuszka, 2003; Harris et al., 1998) |
| Phase separation | • Polymer solution is quenched.<br>• Undergoes a liquid–liquid phase separation to form two phases: a polymer-rich phase and a polymer-poor phase.<br>• The polymer-rich phase solidifies and the polymer-poor phase is removed, resulting in a porous polymer network.<br>(Mikos et al., 2004) |
| Electrospinning | • A force of gravity or mechanical pressure combined with an electric field of high voltage (10–20 kV) is applied to a polymer solution that is pumped to a spinneret.<br>• When the electric charge overcomes the surface tension of the polymer solution, a polymer jet is ejected.<br>• This is followed by solvent evaporation, which forms the solid nanofibers.<br>(Ma et al., 2005) |
| Fiber bonding | • Polymer fibers can bonded at their crosspoints using a secondary polymer, which results in a 3-D porous matrix.<br>(Chung and Park, 2007) |
| Rapid prototyping (RP) | • Uses a 3-D computer-aided design (CAD) model.<br>• Generates components by ink-jet printing onto a polymer powder layer.<br>• The position of the jet is controlled via a CAD program.<br>• Salt can be added into the polymer powder layer to generate pores.<br>(Yang et al., 2002) |

*(Continued)*

| Table 2.3 (Continued) | |
|---|---|
| **Fabrication Techniques** | |
| Melt molding (MM) | • Involves heating the polymer above glass-transition temperature or melting temperature. <br>• The polymers can then be molded into shape. <br>(Oh et al., 2003) |
| Membrane lamination (ML) | • Involves preparation of polymeric membranes followed by stacking of the membranes. <br>• The properties of individual membranes determine bulk properties of the final scaffolds. <br>(Ramalingam et al., 2012) |
| Emulsion freezing/freeze drying (EF/FD) | • Involves homogenization of a polymer solvent solution with water to create an emulsion. <br>• The emulsion is then rapidly cooled to lock the liquid-state structure. <br>• The solvent and water are removed by freeze-drying. <br>(Whang et al., 1995) |

## 2.3 BIOREACTORS

A TE bioreactor can be defined as a device that uses mechanical means to influence biological processes under closely monitored operating conditions (Darling and Athanasiou, 2003). Bioreactors can be used to increase efficiency in cell cultures and reduce the duration of the process. They can provide signals to cells and encourage them to differentiate (Partap et al., 2010).

## 2.4 CELLS

Cell source is a major issue for TE and regenerative medicine. Cells used in TE can be classified into three main types: autologous (patient's own cells), allogenic (human cells obtained from a donor), and xenogenic (cells of animal origin).

Allogeneic tissue has been routinely used for transplantation due to the development of immunosuppressive drug therapies. The furthest advances in the use of allogenic tissues are in the area of connective tissue replacement, cartilage, and skin (Peretti et al., 2001; Eaglstein and Falanga, 1998). Research in the area of xenotransplantation (cross-species transplantation) has grown tremendously over the last few decades. Xenotransplantation has been used clinically in various specialties such as injection of porcine islet cells of Langerhans into patients with type 1 diabetes mellitus (Rood and Cooper, 2006), and transplantation of pig neuronal cells into patients with Parkinson's disease and Huntington's disease (Fink et al., 2000). There are numerous limitations of xenotransplantation, including ethical issues such as animals not being able to consent to such procedures. There is also the risk of transmission of viruses from xenografts into humans (Sobbrio and Jorqui, 2014).

### Autologous cell therapy

Autologous cells harvested from an individual are cultured and reintroduced into the body at the damaged site to repair the tissue. Autologous cells are ideal as they obviate the need for immunosuppressive therapy. However, this strategy has a number of limitations, including the difficulty in harvesting a sufficient number of cells. Therefore, attention has focused on the use of stem cells, such as embryonic stem cells (ESCs) and mesenchymal stem cells (MSCs) (Howard et al., 2008). Stem cells can be categorized according to their differentiation potential (Table 2.4).

### Embryonic stem cells

Human ESCs have the ability to self-renew and to differentiate into many specialized cell types (Brivanlou et al., 2003). Pluripotent cells can be isolated from the inner cell mass of the embryo during the blastocyst stage. However, there are ethical and religious concerns associated with ESCs because embryos are destroyed in order to obtain them (Olson et al., 2011).

### Induced pluripotent stem cells

Induced pluripotent stem cells (iPSCs) hold great promise for cell therapies and TE. An exciting breakthrough in stem cell biology is that adult somatic cells (eg, skin fibroblasts) can be reprogrammed into iPSCs

| Table 2.4 Categories of Stem Cells (Thomson et al., 1998; Verfaillie, 2002) | |
| --- | --- |
| **Types of Stem Cells** | |
| Totipotent | Capable of differentiating into all the tissues that form the human body. |
| Pluripotent (eg, embryonic stem cells) | Present in the inner cell mass of a blastocyst and capable of differentiating into any of the three germ layers (ectoderm, mesoderm, and endoderm). |
| Multipotent (eg, mesenchymal stem cells) | Differentiate into various cell types of a single germ layer. |
| Oligopotent | Differentiate into few cells of a single germ layer. |
| Unipotent | Differentiate into only one type of cell of a single germ layer. |

(Takahashi and Yamanaka, 2006). The iPSCs derived from somatic cells make it possible to design patient-specific cell therapies, which bypass immune rejection issues and the ethical concerns of deriving and using ESCs as a cell source. The unlimited expansion potential of iPSCs also makes them a valuable cell source for TE. However, limited understanding of the mechanism underlying it, such as the appropriate differentiation stage of the cells for specific TE applications, currently limits the clinical applicability of the technique. Nevertheless the future potential of reprogramming is promising (Olson et al., 2011).

**Adult stem cells**

The use of adult stem cells (ASCs) avoids many controversies associated with the use of ESCs. They do not require the use of human embryos for their isolation, and do not form teratomas if injected in vivo. Adults stem cells have been found in many adult tissues, including bone marrow (BM), brain, skin, muscle, and in specific organs. One type of ASC—which has the capacity to differentiate into many different types of tissues, such as osteoblasts, chondrocytes, adipocytes, myocytes, and tenocytes—is the MSC, also known as the multipotent adult progenitor cell. MSCs are nonhematopoietic stromal cells. They were first isolated from BM, and subsequently from other tissues such as adipose tissue. Adipose tissue is a useful source of these cells, as it is more readily accessible than BM-derived MSCs (Pittenger et al., 1999; Pountos and Giannoudis, 2005; Schäffler and Büchler, 2007).

## 2.5 GROWTH FACTORS

Growth factors are soluble signaling molecules that control cellular responses through specific binding of transmembrane receptors on target cells. Growth factors applied to a cell–scaffold construct can help promote tissue regeneration in comparison to non-use of growth factors (Ikada, 2006). Growth factors that have been used in TE include bone morphogenetic proteins, basic fibroblast growth factor (bFGF or FGF-2), vascular epithelial growth factor (VEGF), and transforming growth factor-$\beta$ (TGF-$\beta$) (Lee et al., 2011).

**What is the rationale for growth factor immobilization?**

Growth factor immobilization prevents the loss of bioactivity due to diffusion seen in conventional delivery of growth factors in the soluble

| Table 2.5 Immobilization of Growth Factor to Scaffold (Vasita and Katti, 2006) | | |
|---|---|---|
| Noncovalent growth factor immobilization | Encapsulation | Physical entrapment of the growth factor in the carrier matrix. Carrier systems used to deliver growth factors by their physical entrapment include polymeric microspheres, liposomes, hydrogels, and foams. |
| | Adsorption of the growth factor on the matrix surface | Physical adsorption of growth factors on the carrier material prevents their denaturation. Growth factors that can naturally bind to the extracellular matrix are desirable. |
| | Ion complexation | Formation of polyionic complexes with each other. A positively charged growth factor can be complexed with negatively charged polymer chains in the carrier matrix. The complexed growth factor will be released from the growth factor−carrier complex if a significant environmental change takes place. |
| Covalent growth factor immobilization | Covalent attachment of proteins to the carrier surface | Growth factor is covalently attached to the carrier matrix. |

form. Immobilization also allows a more controlled release of the growth factor.

**How are growth factors immobilized?**

The growth factor can be immobilized to the scaffold by either noncovalent or covalent binding (Table 2.5, Vasita and Katti, 2006).

## 2.6 VASCULARIZATION

A major prerequisite for long-term function of TE constructs is the establishment of an efficient vascularization of the construct after implantation to meet their oxygen and nutrient demands. Over the last few years, two principal vascularization strategies have emerged in the field of TE.

In the first strategy, vascularization can be achieved by stimulating angiogenesis in a number of ways, allowing ingrowth of newly formed blood vessels into implanted tissue constructs from the surrounding host tissue. However, complete vascularization of implants using this strategy is a time-consuming process and is susceptible to tissue loss due to hypoxia. This is due to the slow growth rate of newly developing vessels (Utzinger et al., 2015; Laschke and Menger, 2016). In an alternative strategy, "prevascularization" may overcome these problems. Here, preformed microvascular networks can be inosculated within the construct prior to implantation. These networks then develop interconnections with the host microvasculature at the implantation site (Laschke and Menger, 2012), or by surgical anastomosis of the blood vessels (Beier et al., 2010; Eweida et al., 2011; Laschke and Menger, 2016).

### 2.6.1 Stimulating Angiogenesis

Angiogenesis is the development of new vessels from preexisting blood vessels (Patan, 2000). There are a number of ways angiogenesis can be stimulated in tissue constructs. One strategy to improve scaffold vascularization is incorporation of angiogenic growth factors into the implants. Growth factors stimulate sprouting during angiogenesis. The majority of these factors are stimulated by hypoxia and hypoxia-inducible factors (HIFs), which induce the potent angiogenesis stimulator VEGF-A. Hypoxia and HIFs stimulate endothelial cell survival, proliferation, and capillary sprouting. However, if hypoxia is prolonged, vascular degeneration occurs, endothelial cells undergo apoptosis, and cell death occurs. HIF-1$\alpha$ stimulates vessel sprouting and neovascularization. A deficiency of HIF-1$\alpha$ can lead to nonhealing wounds, but prolonged overexpression can induce hypertrophic scars. HIF-2$\alpha$ promotes vessel remodeling into mature vessels, but overexpression impairs wound healing. Modulation of HIF-1$\alpha$ or HIF-2$\alpha$ expression may therefore be a promising therapeutic strategy to incorporate into scaffold construction (Nauta et al., 2014). Other growth factors which can be used to stimulate angiogenesis are FGFs, platelet-derived growth factors (PDGFs), and TGFs. These factors may originate from cells of the host tissue itself due to tissue injury or an inflammatory process during the implantation procedure. Improved vascularization of scaffolds can also be achieved by seeding them with differentiated tissue-specific cells or stem cells. Cells inside a construct can stimulate the ingrowth of blood vessels by releasing angiogenic growth factors. Stem cells can also differentiate into vascular cells and can self-assemble into microvessels (Laschke and Menger, 2012).

### 2.6.2 Prevascularization

There are a number of current in vitro and in vivo prevascularization approaches used to increase vascularization of TE constructs (Table 2.6).

| Table 2.6 Prevascularization Techniques (Laschke and Menger, 2016) | |
|---|---|
| **In Vitro Prevascularization Approaches** | |
| Scaffold cell seeding | Scaffolds can be seeded with cells such as endothelial cells and mural cells to improve development of newly formed microvessels. (Mural cells are vascular smooth muscle cells and pericytes, both involved in the formation of normal vasculature). |
| Spheroids | Round-shaped clusters of cells that provide a three-dimensional environment with direct cell—cell contacts. |
| Cell sheet technology | These are prevascularized tissue constructs that can be seeded with appropriate cells without the need for scaffold materials. |
| **In Situ Growth Prevascularization Approaches** | |
| Angiogenic ingrowth | This uses the body as a natural bioreactor for the generation of functional microvessels within tissue constructs. The scaffold can be implanted into a well-vascularized and easily accessible tissue of the body. This induces an angiogenic tissue response. After complete vascularization, the implant is excised and transferred to the defect site. |
| Flap technique (Prelamination) | A tissue construct is implanted into a muscle flap to allow ingrowth of newly developing microvessels from the surrounding tissue. The flap is transferred to a tissue defect, where the vascular pedicle of the flap is surgically anastomosed to the blood vessels of the defect site. |
| Arterio-Venous (AV) loop | This strategy forms a shunt loop between an artery and vein. It generates blood vessels by sprouting in the scaffold. The prevascularized tissue construct can then be surgically anastomosed to the blood vessels of the defect site. |

The successful transfer of these approaches into a clinical routine represents one of the major challenges in TE. It requires thorough preclinical and clinical analyses to optimize the efficiency and safety of the different approaches and to also create the most cost-effective option (Laschke and Menger, 2016).

## 2.7 SUMMARY

TE aims to replace biological tissues and organs that have been damaged by trauma or disease. It is a complex scientific field that has the potential to change the delivery of regenerative medicine solutions. The core components of TE in all fields include a scaffold (analogous to the ECM) to recruit and guide host cells to regenerate a tissue, delivery of cells and/or growth factors into the damaged area, and cultivation of cells on the scaffold in a culture system (bioreactor). The tissue construct is implanted into the host, where further maturation and integration are anticipated. The TE paradigm is applicable to a number of systems in the body. Using the core components of TE has enabled the development of novel treatment methods for various diseases where traditional therapies have failed. Currently, TE remains largely experimental in many disease processes. One limiting factor that has hindered the progress of various TE constructs is the lack of vascularization of constructs. It is anticipated that with advances in vascularization techniques, TE can progress further.

## REFERENCES

Adamowicz, J., Kowalczyk, T., Drewa, T., 2013. Tissue engineering of urinary bladder – current state of art and future perspectives. Cent. Eur. J. Urol. 66 (2), 202–206.

Beier, J.P., Horch, R.E., Hess, A., Arkudas, A., Heinrich, J., Loew, J., 2010. Axial vascularization of a large volume calcium phosphate ceramic bone substitute in the sheep AV loop model. J. Tissue Eng. Regen. Med. 4, 216–223.

Brivanlou, A.H., Gage, F.H., Jaenisch, R., Jessell, T., Melton, D., Rossant, J., 2003. Stem cells. Setting standards for human embryonic stem cells. Science 300, 913–916.

Chan, B.P., Leong, K.W., 2008. Scaffolding in tissue engineering: general approaches and tissue-specific considerations. Eur. Spine J. 4, 467–479.

Chung, H.J., Park, T.G., 2007. Surface engineered and drug releasing pre-fabricated scaffolds for tissue engineering. Adv. Drug Deliv. Rev. 59 (4–5), 249–262.

Darling, E.M., Athanasiou, K.A., 2003. Biomechanical strategies for articular cartilage regeneration. Ann. Biomed. Eng. 31, 1114–1124.

Eaglstein, W.H., Falanga, V., 1998. Tissue engineering and the development of Apligraf a human skin equivalent. Adv. Wound Care 11, 1–8.

Eweida, A.M., Nabawi, A.S., Marei, M.K., Khalil, M.R., Elhammady, H.A., 2011. Mandibular reconstruction using an axially vascularized tissue-engineered construct. Ann. Surg. Innov. Res. 5, 2.

Fink, J.S., Schumacher, J.M., Ellias, S.L., 2000. Porcine xenografts in Parkinson's disease and Huntington's disease patients: preliminary results. Cell Transplant 9 (2), 273–278.

Grayson, W.L., Martens, T.P., Eng, G.M., Radisic, M., Vunjak-Novakovic, G., 2009. Biomimetic Approach to Tissue Engineering. Semin. Cell Dev. Biol. 20 (6), 665–673.

Harris, L.D., Kim, B., Mooney, D.J., 1998. Open pore biodegradable matrices formed with gas foaming. J. Biomed. Mater. Res. 42, 396–402.

Hendow, E.K., Guhmann, P., Wright, B., Sofokleous, P., Parmar, N., Day, R.M., 2016. Biomaterials for hollow organ tissue engineering. Fibrogenesis & Tissue Repair 9, 3.

Howard, D., Buttery, L.D., Shakesheff, K.M., Roberts, S.J., 2008. Tissue engineering: strategies, stem cells and scaffolds. J. Anat. 213, 66–72.

Ikada, Y., 2006. Challenges in tissue engineering. J. R. Soc. Interface 3 (10), 589–601.

Jakab, K., Marga, F., Norotte, C., Murphy, K., Vunjak-Novakovic, G., Forgacs, G., 2010. Tissue engineering by self-assembly and bio-printing of living cells. Biofabrication 2 (2), 022001.

Laschke, M.W., Menger, M.D., 2012. Vascularization in tissue engineering: angiogenesis versus inosculation. Eur. Surg. Res. 48, 85–92.

Laschke, M.W., Menger, M.D., 2016. Prevascularization in tissue engineering: current concepts and future directions. Biotechnol. Adv. 34 (2), 112–121.

Lee, K., Silva, E.A., Mooney, D.J., 2011. Growth factor delivery-based tissue engineering: general approaches and a review of recent developments. J. R. Soc. Interface 8 (55), 153–170.

Ma, Z., Kotaki, M., Inai, R., Ramakrishna, S., 2005. Potential of nanofiber matrix as tissue-engineering scaffolds. Tissue Eng. 11, 101–109.

Mikos, A.G., Thorsen, A.J., Czerwonka, L.A., Bao, Y., Langer, R., Winslow, D.N., et al., 1994. Preparation and characterization of poly(L-lactic acid) foams. Polymer 35, 1068–1077.

Mikos, A.G., Lu, L., Temenoff, J.S., Temmser, J.K., 2004. Synthetic bioresorbable polymer scaffolds. In: Ratner, B.D., Hoffman, A.S., Schoen, F.J., Lemons, J.E. (Eds.), An Introduction to Material in Medicine. Elsevier Academic Press, USA, p. 743.

Nair, L.S., Laurencin, C.T., 2007. Biodegradable polymers as biomaterials. Prog. Polym. Sci. 32 (8–9), 762–798.

Nauta, T.D., van Hinsbergh, V.W.M., Koolwijk, P., 2014. Hypoxic Signaling During Tissue Repair and Regenerative Medicine. Day RM, ed. International Journal of Molecular Sciences. 15 (11), 19791–19815.

Norotte, C., Marga, F., Niklason, L., Forgacs, G., 2009. Scaffold-free vascular tissue engineering using bioprinting. Biomaterials 30 (30), 5910–5917.

O'Brien, F., 2011. Biomaterials & scaffolds for tissue engineering. Mater. Today 14 (3), 88–95.

Oh, S.H., Kang, S.G., Kim, E., 2003. Fabrication and characterization of hydrophilic poly(lactic co-glycolic acid)/poly(vinyl alcohol) blend cell scaffolds by melt molding particulate-leaching method. Biomaterials 24 (22), 4011–4021.

Olson, J.L., Atala, A., Yoo, J.J., 2011. Tissue engineering: current strategies and future directions. Chonnam Med. J. 47 (1), 1–13.

Partap, S., Plunkett, N.A., O'Brien, F.J., 2010. Bioreactors in tissue engineering. Tissue Eng.

Patan, S., 2000. Vasculogenesis and angiogenesis as mechanisms of vascular network formation, growth and remodeling. J. Neurooncol. 50, 1–15.

Peretti, G.M., Caruso, E.M., Randolph, M.A., Zaleske, D.J., 2001. Meniscal repair using engineered tissue. J. Orthop. Res. 19, 278–285.

Pittenger, M.F., Mackay, A.M., Beck, S.C., 1999. Multilineage potential of adult human mesenchymal stem cells. Science 5411, 43–147.

Pountos, I., Giannoudis, P.V., 2005. Biology of mesenchymal stem cells. Injury 36, S8–S12.

Ramalingam, M., Haidar, Z., Ramakrishna, S., Kobayashi, H., Haikel, Y., 2012. Integrated Biomaterials in Tissue Engineering. Wiley, USA.

Rood, P.P., Cooper, D.K., 2006. Islet xenotransplantation: are we really ready for clinical trials? Am. J. Transplant 6 (6), 1269–1274.

Sachlos, E., Czernuszka, J.T., 2003. Making tissue engineering scaffolds work. Review on the application of solid free from fabrication technology to the production of tissue engineering scaffolds. Eur. Cells Mater. 5, 29–40.

Schäffler, A., Büchler, C., 2007. Concise review: adipose tissue-derived stromal cells—basic and clinical implications for novel cell-based therapies. Stem Cells 25 (4), 818–827.

Shi, L.B., Cai, H.X., Chen, L.K., Wu, Y., Zhu, S.A., Gong, X.N., et al., 2014. Tissue engineered bulking agent with adipose-derived stem cells and silk fibroin microspheres for the treatment of intrinsic urethral sphincter deficiency. Biomaterials 35 (5), 1519–1530.

Sobbrio, P., Jorqui, M., 2014. An overview of the role of society and risk in xenotransplantation. Xenotransplantation 21 (6), 523–532.

Takahashi, K., Yamanaka, S., 2006. Induction of pluripotent stem cells from mouse embryonic and adult fibroblast cultures by defined factors. Cell 126, 663–676.

Thomson, J.A., Itskovitz-Eldor, J., Shapiro, S.S., Waknitz, M.A., Swiergiel, J.J., Marshal, V.S., et al., 1998. Embryonic stem cell lines derived from human blastocysts. Science 282, 1145−1147.

Utzinger, U., Baggett, B., Weiss, J.A., Hoying, J.B., Edgar, L.T., 2015. Large-scale time series microscopy of neovessel growth during angiogenesis. Angiogenesis 18, 219−232.

Vasita, R., Katti, D.S., 2006. Growth factor-delivery systems for tissue engineering: a materials perspective. Expert Rev. Med. Devices 3 (1), 29−47.

Verfaillie, C.M., 2002. Adult stem cells: assessing the case for pluripotency. Trends Cell Biol. 12, 502−508.

Whang, K., Thomas, C.H., Healy, K.E., 1995. A novel method to fabricate bioabsorbable scaffolds. Polymer 36, 837−842.

Yang, S., Leong, K., Du, Z., Chua, C., 2002. The design of scaffolds for use in tissue engineering. Part II. Rapid prototyping techniques. Tissue Eng. 8, 1−11.

# CHAPTER 3

# Skin Engineering

## F. Akter[1], J. Ibanez[2], and N. Bulstrode[3]

[1]University of Cambridge, Cambridge, United Kingdom [2]Guy's and St Thomas' Hospital, London, United Kingdom [3]Great Ormond Street Hospital, London, United Kingdom

## 3.1 INTRODUCTION

Tissue engineering of skin (STE) has advanced greatly over the last 30 years. There is a vast array of skin substitutes now available commercially. The skin is the largest organ in mammals and has numerous functions in the body, such as thermoregulation, barrier protection, and sensation. The need for skin substitutes is thus significant, particularly for large defects from burns, congenital diseases, trauma, and infections. The treatment of wounds and associated complications exceeds $20 billion annually in the United States (Braddock et al., 1999). Chronic wounds affect 1% of the population at any given time (Crovetti et al., 2004), and often require multiple treatments.

Conventional methods of skin replacement using autologous tissue such as split skin grafts (SSGs) and full-thickness grafts result in significant donor site morbidity. An alternative source of tissue is thus needed. In STE, various biological and synthetic materials are combined with in vitro-cultured cells to generate functional tissues, which can obviate the problems associated with using skin grafts.

## 3.2 ANATOMY OF SKIN

Human skin consists of three layers: the epidermis, dermis, and the hypodermis (Fig. 3.1). The epidermis is the outermost layer of skin. Keratinocytes and melanocytes are prominent cell types of the epidermis. The basement membrane physically separates the epidermis from the underlying dermis. Progenitor cells are located on the basement membrane and undergo continuous self-renewal and differentiation into keratinocytes. The dermis, beneath the epidermis, contains tough connective tissue, hair follicles, and sebaceous and sweat glands.

Tissue Engineering Made Easy. DOI: http://dx.doi.org/10.1016/B978-0-12-805361-4.00003-5

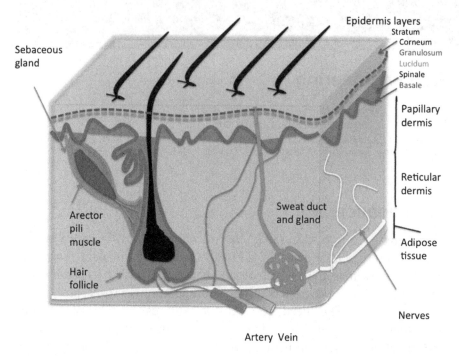

*Figure 3.1 Anatomy of the skin.*

Cellular components of the dermis include the fibroblasts. The deeper subcutaneous tissue (hypodermis) is made of fat and connective tissue (Chen et al., 2009; Pavelka and Roth, 2010).

## 3.3 WOUND HEALING

Wound healing is a multistep process (Table 3.1) that begins with hemostasis after initial damage, followed by inflammation involving recruitment of neutrophils and macrophages, formation of granulation tissue, and tissue remodeling (Chen et al., 2009).

## 3.4 CONVENTIONAL STRATEGIES FOR WOUND DEFECTS

### 3.4.1 Autologous Skin Transplantation

Autologous SSGs are harvested with a dermatome (Fig. 3.2), a tool that is used to remove a layer of epidermis and a superficial part of the dermis. The graft is meshed to increase the surface area and thus reduce the amount of donor tissue needed. The SSG is then applied to the defect site. The donor site usually heals within 7 days;

## Table 3.1 Stages of Wound Healing

**Wound Healing Steps**

| | |
|---|---|
| Hemostasis | Immediately after injury, coagulation and hemostasis take place in the wound (Ross, 1969) |
| Inflammation | During the early inflammatory phase, the complement cascade is activated, which initiates molecular events, leading to infiltration of the wound site by neutrophils within 24–36 h of injury. In the late inflammatory phase, 48–72 h after injury, macrophages appear in the wound and continue the process of phagocytosis (Eming et al., 2007) |
| | The lymphocyte is the last cell type to enter the wound during the inflammatory phase (>72 h after wounding). It appears to be required for wound healing, but with a role that has yet to be defined (Boyce et al., 2000) |
| Proliferation | This process begins at about day 3 and lasts for 2 weeks after wounding. The main processes during the proliferative phase include formation of the ECM, reepithelialization, angiogenesis, and formation of granulation tissue |
| | Reepithelialization<br>Fibroblasts proliferate and migrate to the wound and produce the matrix proteins fibronectin, hyaluronan, collagen, and proteoglycans. Fibroblasts start to construct the new ECM, which supports further ingrowth of cells (Chen et al., 2009). The reepithelialization process is ensured by local keratinocytes at the wound edges and by epithelial stem cells from hair follicles or sweat glands (Reinke and Sorg, 2012) |
| | Angiogenesis<br>The process of forming new blood vessels occurs concurrently during all stages of the healing process (Velnar et al., 2009). The first step in new vessel formation is the binding of growth factors to their receptors on the endothelial cells of existing vessels, thereby activating intracellular signaling cascades. During wound healing, angiogenic capillary sprouts invade the wound clot and organize into a microvascular network (Reinke and Sorg, 2012) |
| | Granulation tissue<br>The last step in the proliferation phase is the development of the acute granulation tissue. It replaces the provisional wound matrix and may produce a scar. It is characterized by a high quantity of cellular compounds and is highly vascular. As a result, it appears with a classic redness and might be traumatized easily (Reinke and Sorg, 2012) |
| Remodeling | Remodeling is the last phase of wound healing and occurs from day 21 up to 1 year after injury, resulting in a mature wound (Reinke and Sorg, 2012) |

however, repeated harvesting at the donor site is associated with delayed healing, pain, scarring, and infection. Full-thickness skin grafts (FTSGs) consist of the entire epidermis and dermis. FTSGs are obtained by careful dissection, and are used for smaller-sized defects to the limited donor sites (Bello et al., 2001; Chen et al., 2009).

## 3.5 TISSUE-ENGINEERED SKIN SUBSTITUTES

Damage to normal skin can disturb its normal vital functions. Bioengineered skin substitutes must provide a number of functions such as preventing fluid loss and contamination. They should also be able to deliver dermal matrix components, cytokines, and growth factors to the wound bed (Shevchenko et al., 2010). The skin substitutes can be classified as acellular versus cellular, epidermal versus dermal, or natural versus synthetic (Table 3.2).

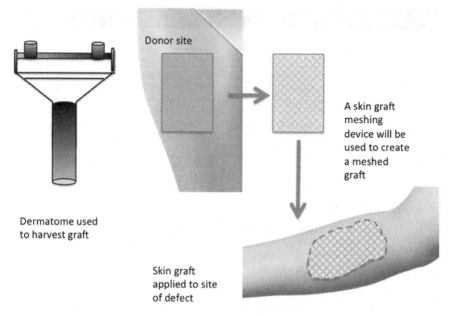

*Figure 3.2 Harvesting and application of split skin graft.*

## 3.5.1 Natural Skin Substitutes

The most widely used biological substitutes worldwide are cadaveric skin allograft, porcine skin xenograft, cultured epithelial autografts (CEAs) and amnion.

### 3.5.1.1 Allograft and Xenograft

Allografts are harvested from consenting donors after death, stored frozen in skin banks, and used when needed until an autograft can be placed onto the wound (Bello et al., 2001). Xenografts made of pig skin have also been used for the same purpose. A major disadvantage of allografts is that they eventually undergo immunogenic rejection, and the site of injury needs to also be covered with an autologous SSG (Chen et al., 2009).

### 3.5.1.2 Cultured Epidermal Autograft (CEA)

Cultured epidermal autograft is an epidermal substitute. A skin biopsy of $2-5\,\text{cm}^2$ is taken from the patient. Keratinocytes purified from the biopsy are grown on mitotically inactivated mouse fibroblasts, isolated, cultured, and expanded into sheets over a period of 3−5 weeks. CEA avoids the mesh aspect of split skin autografts and discomfort at the donor site after skin harvesting. It is, however, limited by the fragility and difficulty of handling, poor attachment rates, and high cost. Commercially available

Table 3.2 Types of Skin Substitutes Available Commercially

| Acellular | Cellular | | Dermal | Epidermal | Natural | Synthetic |
|---|---|---|---|---|---|---|
| | Allogenic Skin Substitutes | Autologous Skin Substitutes | | | | |
| Biobrane® | Apligraf® | Cultured epidermal autograft (CEA) | Integra® | Cultured epithelial autografts (eg, Epicel™, EpiDex®, MySkin®) | Cadaveric skin allograft | Integra® |
| Integra® | Dermagraft® | Cultured skin substitutes (ie, epidermal–dermal substitute) | AlloDerm® | Laserskin (Vivoderm) Autologous keratinocytes on a laser microperforated hyaluronic acid membrane | Porcine skin xenograft | Apligraf® |
| AlloDerm® | TransCyte® | | Dermagraft® | BioSeed-S® Autologous keratinocytes suspended in a fibrin sealant | Amnion | Dermagraft® |
| | | | MatriDerm® | | Cultured epithelial autografts | MatriDerm® |
| | | | | | | Biobrane® |
| | | | | | | OrCel® |
| | | | | | | Hyalomatrix® |
| | | | | | | Renoskin® |

epidermal substitutes include Epicel™, EpiDex®, and MySkin® (Halim et al., 2010).

### 3.5.1.3 Amnion
The amnion is the innermost membrane that encloses the embryo. It can be harvested immediately after birth and preserved to prolong its half-life. It has been used as an effective biological skin substitute in wounds since 1910. It is primarily used for partial thickness burns and temporary coverage of wound. It has been shown to accelerate wound healing, prevent infection, and alleviate pain. It is also thin, comfortable for the patient, and can be easily removed (Halim et al., 2010; Bujang-Safawai et al., 2010).

## 3.5.2 Synthetic Substitutes
Synthetic skin substitutes are constructed out of nonbiological molecules and polymers that are not present in normal skin (Halim et al., 2010). These substitutes should be biodegradable, but should maintain a three-dimensional structure long enough to allow ingrowths of blood vessels, fibroblasts, and coverage by epithelial cells. There are several synthetic skin substitutes available for wound coverage commercially. These include Biobrane®, Dermagraft®, Integra®, Apligraf®, MatriDerm®, OrCel®, Hyalomatrix®, and Renoskin® (Halim et al., 2010).

## 3.5.3 Epidermal Substitutes
See Section 3.5.1.2 describing CEA.

## 3.5.4 Dermal Substitutes
The main component of the dermis is the extracellular matrix (ECM), which provides support for different cell types and skin appendages such as hair, sebaceous glands, and sweat glands. Several scaffolds have been developed to create an environment that mimics the dermal ECM. The biomaterial used in the dermal portion can be natural, synthetic, or both, and made of autologous or allogenic fibroblasts (Halim et al., 2010). Commercially available dermal substitutes include Integra®, AlloDerm®, Dermagraft®, and MatriDerm® (Table 3.3).

## 3.5.5 Epidermal/Dermal Substitutes
Epidermal/dermal substitutes are double-layered skin substitutes. They usually comprise a combination of human allogeneic neonatal keratinocytes and fibroblasts with a scaffold to form a wound cover. There

**Table 3.3 Dermal Substitutes (Shevchenko et al., 2010; Enoch and Kamolz, 2012; Hart et al., 2012)**

| Dermal Substitutes | |
|---|---|
| Integra® | Integra® consists of a matrix of cross-linked collagen and a silicon top layer to mimic the epidermis. It also contains chondroitin 6-sulfate, a glycosaminoglycan that provides elasticity and tensile strength to the scaffold<br>The porous structure of the scaffold allows the host's cells to invade and proliferate within it, thus promoting dermal regeneration while inhibiting wound contraction<br>Integra® is applied using a two-step procedure<br>Step 1: Integra® applied<br>Step 2: The silicon top layer is removed when the artificial dermis appears revascularized<br>Autologous meshed split skin autograft is applied on top of the substitute |
| AlloDerm® | AlloDerm® is an allograft human dermis, aseptically processed to remove cells and freeze-dried to remove moisture while preserving biologic components and the structure of the dermal matrix |
| Dermagraft® | Dermagraft® is a human fibroblast-derived dermal substitute generated by the culture of neonatal dermal fibroblasts onto a bioabsorbable polyglactin mesh scaffold. The human fibroblasts proliferate and secrete collagen, other extracellular matrix proteins, growth factors, and cytokines to create a three-dimensional human tissue containing metabolically active living cells |
| MatriDerm® | This scaffold is composed of a porous collagen scaffold, with elastin hydrolysate that is converted into native host collagen within weeks following application<br>MatriDerm® can be applied straight from the pack onto the wound bed in a single step. The STSG can then be applied. However, there are reports that dermal substituting one-step grafting may affect the survival of an overlying epidermal transplant<br>In common with many other similar products, randomized controlled trials examining the effectiveness of MatriDerm® are currently lacking, making it difficult to safely recommend its widespread use in clinical practice |

**Table 3.4 Commercially Available Epidermal/Dermal Substitutes**

| |
|---|
| Apligraf® is a bilayered bioengineered skin substitute. The US Food and Drug Administration approved it for clinical use in 1998 as the first true composite skin graft. It has been particularly useful in healing diabetic foot and venous leg ulcers that have failed to heal after 3–4 weeks (Zaulyanov and Kirsner, 2007). It is made by culturing human foreskin-derived neonatal fibroblasts in a bovine type I collagen matrix over which human foreskin-derived neonatal epidermal keratinocytes are then cultured and allowed to stratify (Pham et al., 2007) |
| OrCel® is a bilayered cellular matrix in which epidermal keratinocytes and dermal fibroblasts are cultured in two layers into a collagen sponge. It is used in the treatment of chronic wounds and skin graft donor sites (Halim et al., 2010) |

are a number of commercially available substitutes such as Apligraf® and OrCel® (Table 3.4). However, they have limited clinical use worldwide due to high production costs and time to create (Halim et al., 2010).

## 3.6 STEM CELLS FOR SKIN REGENERATION

There are a number of stem cells found in the skin. These include epidermal stem cells and hair follicle stem cells. These stem cells ensure the maintenance of adult skin homeostasis and hair regeneration, and also participate in the repair of the epidermis after injuries. In humans, the first experiments suggesting the existence of epidermal stem cells came from in vitro cultures, where the epithelial cells taken from skin biopsies were multiplied to grow epithelial sheets in the laboratory (Rheinwald and Green, 1975). These epithelial sheets were successfully transplanted to burn victims (Gallico et al., 1984). However, only the epidermis can be replaced with this method, and the new skin has no hair follicles, sweat glands, or sebaceous glands.

The hair follicle is composed of an outer root sheath, an inner root sheath, and the hair shaft. A specialized region of the outer root sheath of the hair follicle, known as the bulge, is a reservoir of stem cells (Tumbar et al., 2004). Stem cells extracted from the human bulge region can differentiate into hair follicle, epidermal, and sebaceous cells in vitro (Roh and Lyle, 2006). In humans, other subpopulations of stem cells reside within the hair follicle, such as melanocyte stem cells, mesenchymal-like stem cells derived from the dermal sheath/dermal papilla, and nestin-positive stem cells (Ojeh et al., 2015).

Several stem cells found in other locations have also been shown to be useful for skin repair. Various studies have shown that mesenchymal stem cells (MSCs) can accelerate wound repair (Fathke et al., 2004; Ishii et al., 2005). Clinical trials have demonstrated that grafts of MSCs promote skin repair both in chronic and acute wounds. This effect has been seen with both the use of bone marrow-derived MSCs (Falanga et al., 2007) and adipose-derived stem cells (Kim et al., 2011).

Although significant advances have been made in STE, there has been limited progress in the development of a full-thickness skin replacement. Poor vascularization of skin grafts is still an unsolved problem, and applying angiogenic growth factors such as vascular endothelial growth factor have not been fully successful due to a short lifespan. The aim is therefore to create a construct containing a time-dependent method of growth factor release. Another limitation that needs to be addressed is that a full-thickness skin replacement must contain skin appendages such as sweat glands and sebaceous glands (Yang and Cotsarelis, 2010). These appendages have limited self-regeneration capability and thus any damage can be permanent. Skin-engineered constructs should also be incorporated with melanocytes to achieve protection from ultraviolet irradiation (Groeber et al., 2011).

An important challenge in skin regeneration is the restoration of sensation. The skin is densely innervated with different types of nerve endings that discriminate between pain, temperature, and sense of touch. When an injury such as a burn occurs, cutaneous nerves and their sensory corpuscles can be destroyed, which causes a patient to complain of loss of skin sensation and chronic cutaneous pain. Cutaneous nerve regeneration can occur from the nerve endings of the wound bed, but it is often limited by scar formation. Tissue-engineered constructs must create a functional ECM to allow nerve migration and prevent scar formation. One strategy that is currently experimental is stem cell differentiation into Schwann cells for better controlled nerve regeneration in the transplant. Tissue-engineered constructs must also incorporate tactile sensors such as sensory corpuscles and hair follicles to create a truly functional replacement for full-thickness wounds (Blais et al., 2013).

## 3.7 SUMMARY

Large, full-thickness wounds need skin grafting to prevent extensive scar formation. Surgical treatment of conditions such as burns is difficult due to limited availability of engraftable skin. Here skin defects can benefit from tissue engineering. Until recently, the gold standard treatment for covering full-thickness skin defects was the use of a skin graft. However, skin grafting can lead to donor site morbidity, and there is limited availability of suitable donor sites. Tissue-engineered skin constructs have been some of the most advanced in the tissue engineering fields, with a number of commercially available products available for wound defects. Currently, however, a full-thickness skin replacement comprising all the layers is not available. A construct that contains all essential components of skin, such as the appendages, will undoubtedly change the field of STE and lead to significant benefits for patients.

## REFERENCES

Bello, Y.M., Falabella, A.F., Eaglstein, W.H., 2001. Tissue-engineered skin. Current status in wound healing. Am. J. Clin. Dermatol. 2 (5), 305–313.

Blais, M., Parenteau-Bareil, R., Cadau, S., Berthod, F., 2013. Concise review: tissue-engineered skin and nerve regeneration in burn treatment. Stem Cells Transl. Med. 2 (7), 545–551.

Boyce, D.E., Jones, W.D., Ruge, F., Harding, K.G., Moore, K., 2000. The role of lymphocytes in human dermal wound healing. Br. J. Dermatol. 143 (1), 59–65.

Braddock, M., Campbell, C.J., Zuder, D., 1999. Current therapies for wound healing: electrical stimulation, biological therapeutics, and the potential for gene therapy. Int. J. Dermatol. 38 (11), 808–817.

Bujang-Safawi, E., Halim, A.S., Khoo, T.L., Dorai, A.A., 2010. Dried irradiated human amniotic membrane as a biological dressing for facial burns: A 7-year case series. Burns 36, 876–882.

Chen, M., Przyborowski, M., Berthiaume, F., 2009. Stem cells for skin tissue engineering and wound healing. Crit. Rev. Biomed. Eng. 37 (4–5), 399–421.

Crovetti, G., Martinelli, G., Issi, M., Barone, M., Guizzardi, M., Campanati, B., et al., 2004. Platelet gel for healing cutaneous chronic wounds. Transfus. Apher. Sci. 30 (2), 145–151.

Eming, S.A., Krieg, T., Davidson, J.M., 2007. Inflammation in wound repair: molecular and cellular mechanisms. J. Invest. Dermatol. 127, 514–525.

Enoch, S., Kamolz, L.P., 2012. Indications for the use of MatriDerm® in the treatment of complex wounds. Wounds Int. 3 (2), 35–41.

Falanga, V., Iwamoto, S., Chartier, M., Yufit, T., Butmarc, J., Kouttab, N., 2007. Autologous bone marrow-derived cultured mesenchymal stem cells delivered in a fibrin spray accelerate healing in murine and human cutaneous wounds. Tissue Eng. 13, 1299–1312.

Fathke, C., Wilson, L., Hutter, J., Kapoor, V., Smith, A., Hocking, A., 2004. Contribution of bone marrow-derived cells to skin: collagen deposition and wound repair. Stem Cells 22, 812–882.

Gallico, G.G., O'Connor, N.E., Compton, C.C., Kehinde, O., Green, H., 1984. Permanent coverage of large burn wounds with autologous cultured human epithelium. N. Engl. J. Med. 311, 448–451.

Groeber, F., Holeiter, M., Hampel, M., Hinderer, S., Schenke-Layland, K., 2011. Skin tissue engineering—in vivo and in vitro applications. Adv. Drug Deliv. Rev. 63 (4–5), 352–366.

Halim, A.S., Khoo, T.L., Mohd Yussof, S.J., 2010. Biologic and synthetic skin substitutes: an overview. Indian J. Plast. Surg. 43 (Suppl.), S23–S28.

Hart, C.E., Loewen-Rodriguez, A., Lessem, J., 2012. Dermagraft: use in the treatment of chronic wounds. Adv. Wound Care 1 (3), 138–141.

Ishii, G., Sangai, T., Sugiyama, K., Ito, T., Hasebe, T., Endoh, Y., 2005. In vivo characterization of bone marrow-derived fibroblasts recruited into fibrotic lesions. Stem Cells 23, 699–706.

Kim, M., Kim, I., Lee, S.K., Bang, S.I., Lim, S.Y., 2011. Clinical trial of autologous differentiated adipocytes from stem cells derived from human adipose tissue. Dermatol. Surg. 37, 750–759.

Ojeh, N., Pastar, I., Tomic-Canic, M., Stojadinovic, O., 2015. Stem cells in skin regeneration, wound healing, and their clinical applications, Blumenberg, M. (Ed.) Int. J. Mol. Sci. 16 (10), 25476–25501.

Pavelka, M., Roth, J., 2010. Functional Ultrastructure: Atlas of Tissue Biology and Pathology, second ed. Springer Science & Business Media, New York, NY.

Pham, C., Greenwood, J., Cleland, H., Woodruff, P., Maddern, G., 2007. Bioengineered skin substitutes for the management of burns: a systematic review. Burns 33 (8), 946–957.

Reinke, J.M., Sorg, H., 2012. Wound repair and regeneration. Eur. Surg. Res. 49, 35–43.

Rheinwald, J.G., Green, H., 1975. Serial cultivation of strains of human epidermal keratinocytes: the formation of keratinizing colonies from single cells. Cell 6, 331–344.

Roh, C., Lyle, S., 2006. Cutaneous stem cells and wound healing. Pediatr. Res. 59 (4 Pt 2), 100R–103R.

Ross, R., 1969. Wound healing. Sci. Am. 220, 40.

Shevchenko, R.V., James, S.L., James, S.E., 2010. A review of tissue-engineered skin bioconstructs available for skin reconstruction. J. R. Soc. Interface 7 (43), 229–258.

Tumbar, T., Guasch, G., Greco, V., Blanpain, C., Lowry, W.E., Rendl, M., 2004. Defining the epithelial stem cell niche in skin. Science 303 (5656), 359–363.

Velnar, T., Bailey, T., Smrkolj, V., 2009. The wound healing process: an overview of the cellular and molecular mechanisms. J. Int. Med. Res. 37 (5), 1528–1542.

WHO, 2014. Burns. World Health Organization, <http://www.who.int/mediacentre/factsheets/fs365/en/>.

Yang, C.C., Cotsarelis, G., 2010. Review of hair follicle dermal cells. J. Dermatol. Sci. 57 (1), 2–11.

Zaulyanov, L., Kirsner, R.S., 2007. A review of a bi-layered living cell treatment (Apligraf®) in the treatment of venous leg ulcers and diabetic foot ulcers. Clin. Interv. Aging 2 (1), 93–98.

# CHAPTER 4

# Neural Tissue Engineering

**F. Akter[1], J. Ibanez[2], and M. Kotter[1]**
[1]University of Cambridge, Cambridge, United Kingdom [2]Guy's and St Thomas' Hospital, London, United Kingdom

## 4.1 INTRODUCTION

The nervous system consists of the central nervous system (CNS) and the peripheral nervous system (PNS). The CNS is made up of the brain and spinal cord, which have limited endogenous regenerative capacity. Neurodegenerative conditions and traumatic injuries to the brain or spinal cord are associated with numerous complications. Following an injury, activated glial cells (including astrocytes and microglia) migrate to the injury site and multiply to create a glial scar (Delcroix et al., 2010). Initially, the glial response helps buffer excitotoxic and cytotoxic molecules, repair the blood—brain barrier (BBB), and isolate the site of injury. However, if the glial cells persist at the injury site, they produce inhibitory factors, which limit axonal regeneration (Shoichet et al., 2008). At present, there is no treatment modality to completely repair the CNS. Current clinical treatment strategies involve stabilization of injuries and alleviation of symptoms with pharmacological agents. However, optimum outcome is seldom achieved, and pharmacological agents have side effects. (Shoichet et al., 2008).

The PNS has some regenerative capacity, but traumatic injuries to nerves usually require surgical treatment. Nerve endings can be sutured directly if the injury affects only a smaller area. Larger defects often need nerve autografts or allografts. Unfortunately, there is usually a limited supply of donor nerves. Moreover, there is the risk of infection and incomplete recovery of nervous tissue (Grinsell and Keating, 2014).

## 4.2 SPINAL CORD INJURY

Spinal cord injury (SCI) defects are complicated, and there is currently no solution to completely repair them. Neural tissue engineering offers

Tissue Engineering Made Easy. DOI: http://dx.doi.org/10.1016/B978-0-12-805361-4.00004-7

hope to patients and is a rapidly growing field that aims to create engineered tissue that can replace and repair damaged tissue. Injury to the spinal cord can result in a permanent disability and is thus of significant psychological, social, and economic morbidity to the patient and their relatives. The annual incidence of SCI in the United States is estimated to be 40 cases per million population (NSCISC, 2014).

A series of pathophysiological steps occurs in SCIs (Fig. 4.1). The first injury is the "primary injury," which causes a series of cellular reactions leading to further damage, called the "secondary injury." Mechanisms involved in secondary injury include microvascular changes, free radical formation, accumulation of excitatory neurotransmitters (eg, glutamate), neural damage due to excessive excitation (excitotoxicity), electrolyte imbalance, inflammatory response, and apoptosis (Samadikuchaksaraei, 2007). Both the primary and secondary injuries lead to loss of neurons, astrocytes, and oligodendrocytes (Rowland et al., 2008). As part of the endogenous repair process following acute injury, there is a migration of cells such as astrocytes,

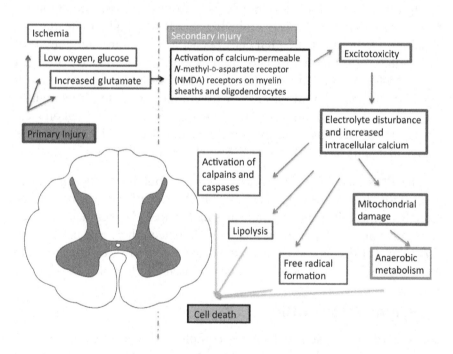

*Figure 4.1 Pathophysiology of spinal cord injury.*

microglia, and Schwann cells (SCs). However, the spinal cord has limited endogenous regenerative capacity (Shoichet et al., 2008). Current treatment strategies, including drug delivery and cell delivery, have been investigated; however, they have been met with variable success. Tissue engineering is an emerging area in biomaterial research that possesses great therapeutic potential. Studies have begun to explore the possibility of using tissue-engineering technology to repair SCIs, specifically by using a biological scaffold, neurotropic growth factors, and appropriate cells.

### 4.2.1 Cell Therapy

As with all forms of tissue engineering, seed cells are needed for application in SCIs. Embryonic stem cells (ESCs) and neural stem cells (NSCs) were used in the early stages of seed cell research. Transplanting ESCs into the brain enables it to differentiate into other cells such as oligodendrocytes and astrocytes, and can improve neurological function. However, this method has been met with ethical issues. Mesenchymal stem cells (MSCs) extracted from bone marrow have the ability to differentiate into various types of brain cells such as those with astrocytic and neuronal phenotypes. However, harvesting MSCs from the bone marrow causes donor site trauma to the patient. Adipose-derived stem cells (ADSCs) have been shown to be a good alternative as adipose tissue is abundant in the body and harvesting them does not create as much trauma to the patient. The cells may differentiate into neurons and can improve neuronal regeneration in SCIs (Ji et al., 2014).

NSCs seeded in a polymer scaffold have shown promise in functional recovery following traumatic SCI (Teng et al., 2002). However, harvesting these cells is difficult. Induced pluripotent stem cells (iPSCs) may be useful in generating these cells. Yuan et al. induced human iPSCs into NSCs. The NSCs were stable, and had the capacity to proliferate and differentiate into neural lineages with no tumor formation (Yuan et al., 2013).

In addition to the seed cells, various other cells—including SCs, macrophages, and dendritic cells (DCs)—are involved in limiting damage to the spinal cord following trauma and in the process of repair after injury. These cells may thus be used in regeneration of spinal tissues.

#### 4.2.1.1 Schwann Cells

SCs are one of the most widely studied cell types for repair of the spinal cord. Following trauma to the spinal cord, SCs migrate from spinal roots into the injury site, where they participate in endogenous repair processes. SC implantations support regeneration and myelination of axons, reduce cyst formation in the injured tissue, and reduce secondary damage of tissue (Williams and Bunge, 2012).

#### 4.2.1.2 Macrophages

After SCI, an inflammatory cascade is initiated. Neutrophils are recruited from the circulation. Recruitment of macrophages is limited in the CNS and the resident microglia cells are the main immune cells that are activated within the first 24 h after SCI (Barami and Diaz, 2000). Two to three days postinjury, blood monocytes migrate to the injury site where they differentiate into macrophages (Donnelly and Popovich, 2008). In the injured spinal cord, macrophages cause an inflammatory phase and limit remodeling. Uncontrolled inflammation is thought to exacerbate the neuronal loss that follows spinal cord trauma. However, controlled inflammation using injected macrophages into contused spinal cords of animal models has been shown to improve motor recovery and reduce spinal cyst formation (Bomstein et al., 2013). Unfortunately, clinical trials have thus far failed to demonstrate a positive neurological response (Lammertse et al., 2012).

#### 4.2.1.3 Dendritic Cells

DCs implanted into the injured spinal cord of mice have been shown to secrete neurotrophic factors such as neurotrophin-3 (NT-3), induce proliferation of endogenous neural stem/progenitor cells, and possibly activate microglia/macrophages, all of which can lead to repair of the injured adult spinal cord (Mikami et al., 2004).

#### 4.2.1.4 Olfactory Ensheathing Cells

Olfactory ensheathing cells (OECs) are specialized glial cells that have properties of both SCs and astrocytes (Gudino-Cabrera and Nieto-Sampedro, 2000). They can be obtained from the olfactory bulb or nasal mucosa (lamina propria) (Samadikuchaksaraei, 2007). OECs transplanted into the demyelinated spinal cord can form compact myelin. They also provide an environment that supports the development of nodes of Ranvier and the restoration of impulse conduction in central demyelinated axons (Sasaki et al., 2006). Transplantation of OECs into animal models of an injured spinal cord results in regrowth of axons across the injury site and recovery of functional behaviors such as walking (Ramon-Cueto et al., 2000). Recent Phase I clinical trials have shown some neurological improvement after implantation of OECs in patients ($n = 3$) with complete SCI. There were no adverse findings related to transplantation of OECs (Tabakow et al., 2013). Further studies with a larger cohort of patients will enable us to determine the extent to which the cell transplants contributed to the neurological improvement seen.

### 4.2.2 Growth Factors

Neurotrophic factors protect neuronal cells from apoptosis and promote axonal regeneration, and thus play an important role in functional recovery following SCI (McCall et al., 2012). Neurotrophic factors include neurotrophins, ciliary neurotropic factor, and the glial cell line-derived neurotrophic factor family. Neurotrophic factors such as neurotrophin nerve growth factor (NGF), NT-3, and brain-derived neurotrophic factor (BDNF) are important for the survival and regeneration of neurons in the brain (Ji et al., 2014). NGF content of the spinal cord is increased after cord injury (Bakhit et al., 1991; Murakami et al., 2002), and therefore has some promise in the treatment of SCI. BDNF-incorporated agarose scaffold implanted into the spinal cord of a rat resulted in the growth of regenerating axons through the scaffold (Stokols et al., 2006). NT-3 has been shown to promote the regenerative growth of the corticospinal tract (CST, the largest tract system leading from the brain to the spinal cord) following SCI (Schnell et al., 1994).

### 4.2.3 Scaffold

To create a neural tissue construct, an appropriate scaffold must be used (Table 4.1). Scaffolds provide a route through which regenerating axons can be guided from one end of the injury to the other, and have potential to promote regeneration in the injured spinal cord. Scaffolds for spinal cord regeneration must be biocompatible, have ideal degradation rates, possess mechanical properties suitable for cell adhesion and axonal regrowth, and be easy to transplant (Samadikuchaksaraei, 2007).

**Table 4.1 Scaffolds Used in Spinal Cord Injury**

| Natural Materials Used as Scaffold Materials | |
|---|---|
| Collagen | Collagen supports neural cell attachment and growth. In SCIs, collagen has been shown to support axonal regeneration (Tsai et al., 2006). Surgical reconstruction of transected cat spinal cord using collagen increased regenerative activity and led to functional recovery (Kataoka et al., 2004). Functional recovery has also been observed after collagen implantation in a patient with SCI (Kataoka et al., 2004). |
| Silk | Silk has been used as a scaffold in SCI experiments. It has good compatibility and induces slight inflammation reaction in vivo. However, silk can be brittle and may need to be combined with another polymer such as chitosan to improve its mechanical properties for use in SCI treatments (Ji et al., 2014). |
| Alginate | Alginate implanted into the spinal cords of rats has been shown to increase axonal elongation (Suzuki et al., 1999). |
| **Synthetic Polymers** | |
| | Synthetic polymers have been shown to repair SCIs in rats. Nanofibrous collagen nerve conduits made of poly (D,L-lactide-co-glycolide) can promote neural fiber growth following SCI, and are also capable of inhibiting glial scar hyperplasia (Liu et al., 2012). Polyethylene glycol, a water-soluble surfactant polymer, has been shown to repair neuronal membrane damage and enhance functional recovery, and thus may be used in a tissue engineering approach to SCI (Shi, 2013). |

## 4.3 BRAIN TISSUE ENGINEERING

Long-term effective treatments for neurodegenerative diseases and brain injury are lacking. The prevalence of neurodegenerative diseases increases every year, partly due to the increase in higher life expectancy. Parkinson's disease (PD) is the most studied neurodegenerative disorder, and affects 2% of the population after 65 years of age. Patients with PD suffer numerous debilitating symptoms. These occur due to the loss of dopaminergic neurons in the substantia nigra pars compacta, which leads to a deficiency of dopamine. The current standard treatment of PD is the drug L-Dopa; however, it has many side effects (Delcroix et al., 2010). Huntington's disease, is another neurodegenerative disorder; it results in neuronal dysfunction and degeneration, and causes numerous psychological and physical symptoms. There is currently no cure, and treatments focus on reducing symptoms (Delcroix et al., 2010). Stroke is a leading cause of serious long-term disability (Mozaffarian et al., 2015), and kills almost 130,000 Americans each year (CDC, 2015). Traumatic brain injury (TBI) is a major source of morbidity to patients and their relatives. It has an annual incidence of over 1.4 million in the U.S. (Delcroix et al., 2010).

Brain tissue engineering strategies are being studied in order to develop alternative treatments to overcome some of the limitations of current treatment strategies. Scaffolds are currently being designed as tissue regenerative implants for damaged brain tissue. Scaffolds can be implanted with neural stem or progenitor cells to increase neuron inter-connectivity and endogenous cell infiltration, and also provide cells with an artificial extracellular matrix (ECM) network (Shoichet et al., 2008).

**What essential properties do scaffolds in brain tissue engineering need to meet?**

- Scaffolds must be biocompatible, biodegradable, and small enough to infiltrate cells and release drugs in a controlled manner.
- Scaffolds must also preserve the integrity of the BBB.
- Scaffolds must be porous, and have optimal pore dimensions and surface functionality, and must support neural cells.
- Scaffolds must not elicit host immune system reactions.

**Which type of scaffolds have been investigated as candidates for neural tissue engineering within the brain?**

A range of scaffolds—including hydrogels, self-assembling peptides, and electrospun nanofiber scaffolds—have been investigated as neural tissue engineering candidates within the brain.

Hydrogels have been shown to encourage cell infiltration, reduce glial scar formation, and promote neurite extension when implanted into a brain lesion (Hou et al., 2005).

**Describe the use of stem cells in brain tissue engineering and their limitations**

Transplanting ESCs into the brain is thought to improve neurological function as they can differentiate into neuronal cells. Human ESCs cultured with a combination of growth factors have been shown to generate myelinating oligodendrocyte precursors (Nistor et al., 2005). Mouse ESCs injected directly into 6-hydroxy-dopamine-treated rat brains (to kill dopamine neurons and induce Parkinson's symptoms) led to relief of Parkinson-like symptoms (Bjorklund et al., 2002). However, this method has certain ethical issues, as well as problems with regard to rejection reactions.

MSCs transplanted into the brain promote endogenous neuronal growth, decrease apoptosis, and regulate inflammation. When transplanted at sites of nerve injury, they have been shown to promote functional recovery by producing trophic factors that induce regeneration of host neurons (Crigler et al., 2006). Transplantation of cells into the brain of immunodeficient mice increased the proliferation of endogenous NSCs (Munoz et al., 2005).

However, MSC injection into the brain carries risks such as the potential for ectopic differentiation to other tissue lineages. Also, MSCs have revascularization capacity, and thus cannot be used in conditions such as brain tumors, where there may increase survival of the tumors (Joyce et al., 2010).

NSCs are multipotent stem cells with the capacity to differentiate into the major cells of the CNS. These cells have been used in combination with biomaterial scaffolds for enhanced delivery of cells following TBI (Tate et al., 2002). NSCs transplanted after experimental TBI have also been shown to promote motor and cognitive recovery (Riess et al., 2002).

## 4.4 TISSUE ENGINEERING OF THE PERIPHERAL NERVOUS SYSTEM

The PNS is composed of two types of cells: neuroglia and neurons. The neuroglia present in the PNS are the SCs and satellite cells. The neurons are nerve cells (Fig. 4.2A) involved in signal transmission between tissues and the CNS. The nerve fibers of the PNS are composed of axons enveloped by SCs, which produce myelin (Birch, 2011). Surrounding the fibers are vessels and connective tissue layers, the endoneurium, the perineurium, and the epineurium (Fig. 4.2B). The endoneurium holds the fibers together to form an endoneurial tube (Thomas, 1963).

### 4.4.1 Peripheral Nervous System Injury

PNS damages are frequent, and result in psychological and physical morbidity for the patient. In the U.S., approximately 2.8% of traumatic injuries affect the PNS (Noble et al., 1998).

During damage to the PNS, if the axon is completely damaged, the two ends of a nerve will retract, causing a gap. The proximal end is the one attached to the cell body, and the distal end is the free-floating

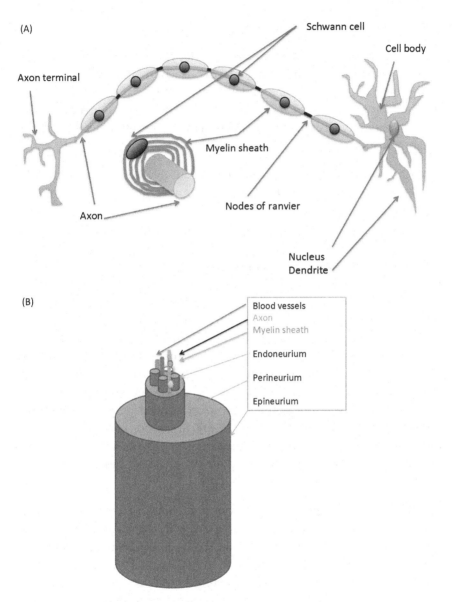

(A)

Schwann cell

Cell body

Axon terminal

Myelin sheath

Nodes of ranvier

Axon

Nucleus
Dendrite

(B)

Blood vessels
Axon
Myelin sheath

Endoneurium

Perineurium

Epineurium

*Figure 4.2 (A) The neuron is composed of the cell body, dendrites, and axon terminals. The nerve impulse begins at the dendrites and is passed to the next nerve through the axon terminals. (B) A nerve fiber with the three layers of connective tissue.*

end. Injury leads to Wallerian degeneration in the distal stump of the injured nerve, and axonal retrograde degeneration near the proximal stump. In Wallerian degeneration, the axons and myelin degenerate completely (Grinsell and Keating, 2014).

**What happens during regeneration?**

There is a migration of macrophages, monocytes, and SCs into the nerve stumps to remove myelin and axon debris (Burnett and Zager, 2004). SCs proliferate, maintain the endoneurium, and produce various neurotrophic factors and ECM molecules to stimulate axon regeneration. They also maintain the endoneurium. New axonal sprouts emerge from the nodes of Ranvier and undergo remyelination by SCs (Gaudet et al., 2011).

In humans, small nerves regenerate at an average rate of 2 mm/day and larger nerves regenerate at an average rate of 5 mm/day (Recknor and Mallapragada, 2006).

## 4.4.2 Common Repair Techniques

Current treatment modalities for PNS injuries include direct suturing of the nerve stumps (neurorrhaphy), if possible. However, when this is not possible, other strategies must be used. The current strategies include: autografting, allografting, and the implantation of nerve guidance conduits (NGCs) to bridge the nerve stumps (Nawrotel, 2015). Autologous nerve grafts are the current gold standard therapy. However, there is a limited supply of donor nerve tissue, two surgical incisions are required, and full functional recovery is often not achieved. Allografts and xenografts can be alternatives to autografts, but their main drawback lies in a possible immune response (Isaacs and Browne, 2014; Saheb-Al-Zamani et al., 2013).

NGCs have the potential to overcome some of these issues. They can be made with a number of different types of scaffolds (Nawrotel, 2015) (Table 4.2). These scaffolds are composed of biodegradable biomaterials, which can integrate with the damaged tissue and remain intact until the nerve fibers have restored connections. The ideal biomaterial should possess properties corresponding to ones of the natural connective tissues of the peripheral nerve. The scaffolds can also be combined with cell-based therapy such as MSCs, which can differentiate in vitro into cells expressing neuronal markers and participate in nerve regeneration (Faroni et al., 2015). The NGCs could also incorporate growth factors (such as NGF) that have been identified during nerve regeneration, cytokines, and other signaling molecules in order to encourage nerve growth migration (Nawrotel, 2015).

| Scaffolds | Cells | Neurotrophic Growth Factors | Extracellular Matrix Molecules to Guide Axons |
|---|---|---|---|
| Collagen | Schwann cells | Nerve growth factor | Laminin |
| Chitosan and silk fibroin | Olfactory ensheathing cells | Glial-derived neurotrophic factor | Fibronection |
| Fibronectin | Embryonic stem cells | Brain-derived neurotrophic factor | |
| | Neural stem cells | Neurotrophin-3 (NT-3) Neurotrophin-4 (NT-4) | |
| | Induced pluripotent stem cells | | |
| | Adipose-derived stem cells | | |
| | Mesenchymal stem cells | | |

Table 4.2 NGC Composition: Scaffolds, Cells, Extracellular Molecules, and Growth Factors

## 4.5 SUMMARY

Neural tissue engineering is one of the most fascinating aspects of tissue engineering. It has tremendous potential in the treatment of various debilitating conditions. One of the problems that the CNS faces is its minimal regenerative capacity. Consequently, patients who suffer injuries or degenerative disease must also cope with significant complications of the disease, without much hope for improvement. Tissue engineering has shown some promise in treating traumatic injuries to the spinal cord, and there have been a number of animal studies and clinical trials investigating the prospect of stem cell therapy, some in combination with the use of biomaterial scaffolds. In the PNS, the likelihood of developing a tissue-engineered construct that can be used clinically is good. Clinical trials using these constructs must be conducted, and a cost-effective method of creating these constructs to reduce the morbidity of using autografts must be sought.

## REFERENCES

Bakhit, C., Armanini, M., Wong, W.L.T., Bennett, G.L., Wrathall, J.R., 1991. Increase in nerve growth factor-like immunoreactivity and decrease in choline acetyltransferase following contusive spinal cord injury. Brain Res. 554, 264–271.

Barami, K., Diaz, F.G., 2000. Cellular transplantation and spinal cord injury. Neurosurgery 47, 691–700.

Birch, R., 2011. Surgical Disorders of the Peripheral Nerve, second ed. Springer, London.

Bjorklund, L.M., Sanchez-Pernaute, R., Chung, S., 2002. Embryonic stem cells develop into functional dopaminergic neurons after transplantation in a Parkinsonian rat model. Proc. Natl. Acad. Sci. U.S.A. 99, 2344–2349.

Bomstein, Y., Marder, J.B., Vitner, K., Smirnov, I., Lisaey, G., Butovsky, O., et al., 2013. Features of skin-coincubated macrophages that promote recovery from spinal cord injury. J. Neuroimmunol. 142, 10−16.

Burnett, M.G., Zager, E.L., 2004. Pathophysiology of peripheral nerve injury: a brief review. Neurosurg. Focus 16, E1.

CDC, NCHS. Underlying Cause of Death 1999−2013. Data are from the Multiple Cause of Death Files, 1999−2013, as compiled from data provided by the 57 vital statistics jurisdictions through the Vital Statistics Cooperative Program. <http://wonder.cdc.gov/ucd-icd10.html>; Accessed April 15, 2016.

Crigler, L., Robey, R.C., Asawachaicharn, A., et al., 2006. Human mesenchymal stem cell subpopulations express a variety of neuroregulatory molecules and promote neuronal cell survival and neuritogenesis. Exp. Neurol. 198, 54−64.

Delcroix, G.J., Schiller, P.C., Benoit, J.P., Montero-Menei, C.N., 2010. Adult cell therapy for brain neuronal damages and the role of tissue engineering. Biomaterials 31 (8), 2105−2120.

Donnelly, D.J., Popovich, P.G., 2008. Inflammation and its role in neuroprotection, axonal regeneration and functional recovery after spinal cord injury. Exp. Neurol. 209, 378−388.

Faroni, A., Mobasseri, S.A., Kingham, P.J., Reid, A.J., 2015. Peripheral nerve regeneration: experimental strategies and future perspectives. Adv. Drug Deliv. Rev. 82−83, 160−167.

Gaudet, A.D., Popovich, P.G., Ramer, M.S., 2011. Wallerian degeneration: gaining perspective on inflammatory events after peripheral nerve injury. J. Neuroinflammation 8, 110.

Grinsell, D., Keating, C.P., 2014. Peripheral nerve reconstruction after injury: a review of clinical and experimental therapies. BioMed. Res. Int. 2014, 698256.

Gudino-Cabrera, G., Nieto-Sampedro, M., 2000. Schwann-like macroglia in adult rat brain. Glia 30, 49−63.

Hou, S., Xu, Q., Tian, W., Cui, F., Cai, Q., Ma, J., et al., 2005. The repair of brain lesion by implantation of hyaluronic acid hydrogels modified with laminin. J. Neurosci. Methods 148 (1), 60−70.

Isaacs, J., Browne, T., 2014. Overcoming short gaps in peripheral nerve repair: conduits and human acellular nerve allograft. Hand 9, 131−137.

Ji, W., Hu, S., Zhou, J., Wang, G., Wang, K., Zhang, Y., 2014. Tissue engineering is a promising method for the repair of spinal cord injuries. Exp. Therap. Med. 7 (3), 523−528.

Joyce, N., Annett, G., Wirthlin, L., Olson, S., Bauer, G., Nolta, J.A., 2010. Mesenchymal stem cells for the treatment of neurodegenerative disease. Regen. Med. 5 (6), 933−946. Available from: http://dx.doi.org/10.2217/rme.10.72.

Kataoka, K., Suzuki, Y., Kitada, M., Hashimoto, T., Chou, H., Bai, H., et al., 2004. Alginate enhances elongation of early regenerating axons in spinal cord of young rats. Tissue Eng. 10, 493−504.

Lammertse, D., Jones, L.A., Charlifue, S.B., Kirshblum, S.C., Apple, D.F., Ragnarsson, K.T., et al., 2012. Autologous incubated macrophage therapy in acute, complete spinal cord injury: results of the phase 2 randomized controlled multicenter trial. Spinal Cord 50 (9), 661−671.

Liu, T., Houle, J.D., Xu, J., Chan, B.P., Chew, S.Y., 2012. Nanofibrous collagen nerve conduits for spinal cord repair. Tissue Eng. Part A 18 (9−10), 1057−1066.

McCall, J., Weidner, N., Blesch, A., 2012. Neurotrophic factors in combinatorial approaches for spinal cord regeneration. Cell Tissue Res. 349, 27−37.

Mikami, Y., Okano, H., Sakaguchi, M., Nakamura, M., Shimazaki, T., Okano, H.J., et al., 2004. Implantation of dendritic cells in injured adult spinal cord results in activation of endogenous neural stem/progenitor cells leading to de novo neurogenesis and functional recovery. J. Neurosci. Res. 76 (4), 453−465.

Mozaffarian, D., Benjamin, E.J., Go, A.S., et al., 2015. Heart disease and stroke statistics—2015 update: a report from the American Heart Association. Circulation; e29–322.

Munoz, J.R., Stoutenger, B.R., Robinson, A.P., et al., 2005. Human stem/progenitor cells from bone marrow promote neurogenesis of endogenous neural stem cells in the hippocampus of mice. Proc. Natl. Acad. Sci. U.S.A. 102, 18171–18176.

Murakami, Y., Furukawa, S., Nitta, A., Furukawa, Y., 2002. Accumulation of nerve growth factor protein at both rostral and caudal stumps in the transected rat spinal cord. J. Neurol. Sci., 198(1-2)63–69.

Nawrotel, K., 2015. Current approaches to peripheral nervous tissue regeneration – mimicking nature. A review. J. Res. Innov. 1, 16–33.

Nistor, G.I., Totoiu, M.O., Haque, N., Carpenter, M.K., Keirstead, H.S., 2005. Human embryonic stem cells differentiate into oligodendrocytes in high purity and myelinate after spinal cord transplantation. Glia 49 (3), 385–396.

Noble, J., Munro, C.A., Prasad, V.S.S.V., Midha, R., 1998. Analysis of upper and lower extremity peripheral nerve injuries in a population of patients with multiple injuries. J. Trauma 45, 116–122.

NSCISC, 2014. Spinal Cord Injury Facts and Figures at a Glance. National Spinal Cord Injury Statistical Center, Birmingham, AL. Available online: <https://www.nscisc.uab.edu/PublicDocuments/fact_figures_docs/Facts%202014.pdf>.

Ramon-Cueto, A., Cordero, M.I., Santos-Benito, F.F., Avila, J., 2000. Functional recovery of paraplegic rats and motor axon regeneration in their spinal cords by olfactory ensheathing glia. Neuron 25, 425–435.

Recknor, J.B., Mallapragada, S.K., 2006. Tissue Engineering and Artificial Organs, third ed. Taylor and Francis Group, Boca Raton, FL.

Riess, P., Zhang, C., Saatman, K.E., Laurer, H.L., Longhi, L.G., Raghpathi, R., et al., 2002. Transplanted neural stem cells survive, differentiate, and improve neurological motor function after experimental traumatic brain injury. Neurosurgery 51 (4), 1043–1052, 1052–54.

Rowland, J.W., Hawryluk, G.W., Kwon, B., Fehlings, M.G., 2008. Current status of acute spinal cord injury pathophysiology and emerging therapies: promise on the horizon. Neurosurg. Focus 25 (5), E2.

Saheb-Al-Zamani, M., Yan, Y., Farber, S.J., Hunter, D.A., Newton, P., Wood, M.D., et al., 2013. Limited regeneration in long acellular nerve allografts is associated with increased Schwann cell senescence. Exp. Neurol. 247, 165–177.

Samadikuchaksaraei, A., 2007. An overview of tissue engineering approaches for management of spinal cord injuries. J. Neuroeng. Rehabil. 4, 15.

Sasaki, M., Black, J.A., Lankford, K.L., Tokuno, H.A., Waxman, S.G., Kocsis, J.D., 2006. Molecular reconstruction of nodes of Ranvier after remyelination by transplanted olfactory ensheathing cells in the demyelinated spinal cord. J. Neurosci 26 (6), 1803–1812.

Schnell, L., Schneider, R., Kolbeck, R., Barde, Y.A., Schwab, M.E., 1994. Neurotrophin-3 enhances sprouting of corticospinal tract during development and after adult spinal cord lesion. Nature 367, 170–173.

Shi, R., 2013. Polyethylene glycol repairs membrane damage and enhances functional recovery: a tissue engineering approach to spinal cord injury. Neurosci. Bull. 29 (4), 460–466.

Shoichet, M.S., Tate, C.C., Baumann, M.D., et al., 2008. Strategies for regeneration and repair in the injured central nervous system. In: Reichert, W.M. (Ed.), Indwelling Neural Implants: Strategies for Contending with the In Vivo Environment. CRC Press/Taylor & Francis, Boca Raton (FL).

Stokols, S., Sakamoto, J., Breckon, C., Holt, T., Weiss, J., Tuszynski, M.H., 2006. Templated agarose scaffolds support linear axonal regeneration. Tissue Eng. 12, 2777–2787.

Suzuki, K., Suzuki, Y., Ohnishi, K., Endo, K., Tanihara, M., Nishimura, Y., 1999. Regeneration of transected spinal cord in young adult rats using freeze-dried alginate gel. Neuroreport 10, 2891–2894.

Tabakow, P., Jarmundowicz, W., Czapiga, B., Fortuna, W., Miedzybrodzki, R., Czyz, M., et al., 2013. Transplantation of autologous olfactory ensheathing cells in complete human spinal cord injury. Cell Transplant. 22 (9), 1591–1612.

Tate, M.C., Hoffman, S.W., Stein, D.G., Archer, D.R., LaPlaca, M.C., 2002. Fibronectin promotes survival and migration of primary neural stem cells transplanted into the traumatically injured mouse brain. Cell Transplant. 11 (3), 283–295.

Teng, Y.D., Lavik, E.B., Qu, X., Jitka, Q., Zurakowski, D., Langer, R., et al., 2002. Functional recovery following traumatic spinal cord injury mediated by a unique polymer scaffold seeded with neural stem cells. Proc. Natl. Acad. Sci. U. S. A. 99 (5), 3024–3029.

Thomas, P.K., 1963. The connective tissue of peripheral nerve: an electron microscope study. J. Anat. 97, 35–44.

Tsai, E.C., Dalton, P.D., Shoichet, M.S., Tator, C.H., 2006. Matrix inclusion within synthetic hydrogel guidance channels improves specific supraspinal and local axonal regeneration after complete spinal cord transection. Biomaterials 27, 519–533.

Williams, R.R., Bunge, M.B., 2012. Schwann cell transplantation: a repair strategy for spinal cord injury? Prog. Brain Res. 201, 295–312.

Yuan, T., Liao, W., Feng, N.H., Lou, Y.L., Niu, X., Zhang, A.J., et al., 2013. Human induced pluripotent stem cell-derived neural stem cells survive, migrate, differentiate, and improve neurological function in a rat model of middle cerebral artery occlusion. Stem Cell Res. Ther. 4, 73.

# CHAPTER 5

# Ophthalmic Tissue Engineering

**F. Akter**
University of Cambridge, Cambridge, United Kingdom

## 5.1 INTRODUCTION

There are approximately 285 million people worldwide who are visually impaired (WHO, 2014). The most common eye conditions are cataracts, glaucoma, macular degeneration, diabetic retinopathy, and corneal diseases. Conventional treatments for eye diseases, such as those affecting the cornea, include the use of grafts. However, there are limitations associated with such procedures, such as rejection of the graft and donor shortage. Consequently, there has been a huge interest and advancement in developing tissue-engineered products (Fagerholm et al., 2009).

In this chapter, the basic anatomy and physiology of the eye will be described, followed by a discussion on the use of tissue-engineered products in ophthalmic diseases.

## 5.2 ANATOMY

The eye is a fluid-filled sphere composed of three layers of tissue (Fig. 5.1) (Purves et al., 2001).

The outer layer is a white, fibrous tissue called the *sclera*. The sclera is covered by the conjunctiva, which continues to cover the inner lining of the eyelids.

The middle layer is called the *uvea*. It contains three continuous components: iris, ciliary body, and choroid.

- The *iris* contains the sphincter pupillae and dilator pupillae muscles, and changes the size of the pupil. The pupil is a transparent structure in the middle of the iris, through which light enters.

Tissue Engineering Made Easy. DOI: http://dx.doi.org/10.1016/B978-0-12-805361-4.00005-9

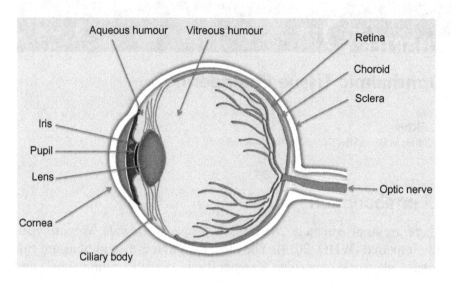

*Figure 5.1 Anatomy of the eye.*

- The *ciliary body* encircles the lens (sits posterior to the iris) and contains the ciliary muscle (adjusts the refractive power of the lens), blood vessels, and fibrous connective tissue.
- The *choroid* is the vascular layer of the eye. It is the main blood supply for the photoreceptors of the retina.
- The *cornea* is the transparent front part of the eye that covers the iris, pupil, and anterior chamber.

  The inner layer is called the *retina*.

- It contains specialized cells that convert visual images into electrical signals. The retina then transmits the signals to the brain via the optic nerve.

## 5.3 OPHTHALMIC TISSUE ENGINEERING

Tissue engineering (TE) holds enormous promise for repairing damage due to diseases of the eye. There is currently a huge effort to develop novel methods to regenerate eye tissues damaged by various diseases, including those affecting the cornea and retina.

## 5.4 CORNEAL DISEASES

Corneal diseases represent a major cause of blindness worldwide. Corneal transplantation advanced rapidly in the last decade with the development of new procedures. The cornea can be transplanted in full (penetrating keratoplasty (PK)), or only partially (lamellar keratoplasty). With the increasing demand for corneal tissue due to an aging population, there is a shortage of donor corneas suitable for transplantation. Progress in TE may thus offer new therapeutic solutions for the replacement of a diseased cornea.

The cornea comprises five layers: the epithelium, Bowman's layer, the stroma, Descemet's membrane, and the endothelium.

## Chambers of Fluid in the Eye

| Anterior chamber | (Between cornea and iris) |
| --- | --- |
| Posterior chamber | (Between iris and lens) |
| Vitreous chamber | (Between the lens and the retina) |

The anterior and posterior chambers are filled with aqueous humor. The vitreous chamber is filled with the vitreous humor.

**Aqueous Humor**
Gelatinous fluid;
Secreted from the ciliary epithelium;
Maintains intraocular pressure.

**Vitreous Humor**
Gelatinous fluid;
Contains many phagocytes to remove cellular debris from visual field;
Avascular (Forrester et al., 2016).

## The Corneal Epithelium

The corneal epithelium is attached to the conjunctival epithelium.
The narrow zone between the cornea and the conjunctiva is known as the limbus. The limbus is the source of corneal epithelial stem cells in humans.
Limbal stem cell proliferation maintains the cornea.
Limbal stem cells prevent conjunctival epithelial cells from migrating onto the corneal surface.
Damage to the limbus leads to conjunctivalization of the cornea (Ezhkova and Fuchs, 2010; Forrester et al., 2016).

The corneal endothelium is the innermost layer of the cornea. The main function of these cells is to pump fluid out of the corneal stroma, allowing the cornea to remain optically clear. Loss of functionality of the endothelial layer results in stromal edema and vision loss. Descemet's stripping automated endothelial keratoplasty (DSAEK) is the gold standard for the surgical treatment of corneal endothelial diseases. It involves removing the endothelium and its underlying basement membrane (Descemet's membrane) and replacing it with a layer of posterior cornea cut from a donor eye. Compared to PK, DSAEK offers many advantages. This procedure promotes faster visual recovery and more predictable refractive outcomes. The main concern regarding the use of DSAEK is a high rate of postoperative endothelial cell loss and a risk of postoperative graft dislocation (Gorovoy, 2006).

**Describe one TE technique to replace diseased corneal endothelium**

Fibroblasts can be cultivated and induced to secrete their own extracellular matrix and form sheets to reconstruct stromal tissue. Corneal endothelial cells can then be seeded on each side of the reconstructed stroma to create a tissue-engineered cornea similar to the native cornea. This method is called the self-assembly approach (Carrier et al., 2009; Jay et al., 2015).

**What are keratoprosthetic devices?**

These are acellular artificial implants consisting of a central optic held in a cylindrical frame (Fig. 5.2). The keratoprosthesis replaces the section of cornea that has been removed. The development of a keratoprosthesis has encountered numerous difficulties, and many of the early corneal transplants had high infection rates and extrusion.

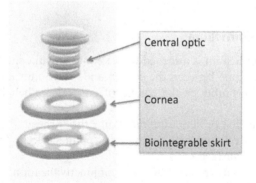

*Figure 5.2 Schematic diagram of a keratoprosthesis.*

In the late 1980s, the concept of a "core-and-skirt" device, in which a biointegrable "skirt" surrounds a central optic, became the most common design. The importance of size and contiguity of pores in the porous skirt were also fully appreciated (Avadhanam et al., 2015; Crawford et al., 2002). Keratoprosthesis such as the AlphaCor, previously known as the Chirila keratoprosthesis, consists of a polymethylmethacrylate device with a central optic region fused with a surrounding sponge skirt. The procedure to implant a keratoprosthesis is, however, complicated, and to date, clinical use of keratoprostheses has been infrequent.

**What are the challenges of corneal TE?**

The structure of the cornea is unique and the collagen architecture is difficult to replicate.

**What are the properties of scaffolds required for corneal TE?**

When constructing a tissue-engineered cornea, the choice of scaffold material is vital to enable the cells to form the same complex arrangement as they would in vivo. Scaffold materials available for corneal TE include acellular corneal stroma and collagen (Table 5.1).

Scaffold properties:

- Biocompatibility;
- Optical transparency;
- Strong enough to withstand handling during surgery;

| Table 5.1 Scaffold Materials Used in Corneal Tissue Engineering (Wang et al., 2013) | |
|---|---|
| Scaffold Materials Used in Corneal Tissue Engineering | |
| Acellular Cornea Stroma | Derived from the acellular allogeneic or autologous graft |
| Amniotic Membrane | Situated in the inner membrane of fetal membranes, including the monolayer of epithelial cells, thick basement membrane, and avascular stroma |
| Collagen | Natural protein material commonly used in TE; Type I collagen is the major composition of human cornea;Collagen fibrils provide physical support to tissues; Can promote cell adhesion and proliferation better than synthetic polymers |
| Fibrin | Commonly used protein; Human fibrin is low in price, readily available, and has good tolerance to cells |
| Chitin | Chitin is a linear polysaccharide; Nontoxic; Promotes growth factor production and acts as a carrier for the release of growth factors |
| Silk fibroin | Extracted from natural silk protein polymer fibers; Transparent, easy to handle, free from disease transmission |

- Flexible enough to take the shape of the eye and lay flat on the surface;
- Easily produced with high speed and low cost;
- Controllable biodegradability or bioresorbability.

## 5.5 RETINAL DISEASES

Genes, lifestyle, and age-related factors can all affect the retina, which is susceptible to numerous disorders. The retina is a light-sensitive multilayered tissue that is actually part of the central nervous system. During development, the retina forms from an outpouching of the diencephalon (posterior part of the forebrain), which invaginates to form the optic cup. The inner wall of the optic cup gives rise to the retina, and the outer wall gives rise to the retinal pigment epithelium (RPE). The RPE is a support tissue for the photoreceptors of the outer retina. Light energy is converted to electrical energy within the photoreceptors. The visual information from the photoreceptors is transmitted to the ganglion cells via intermediate neurons (Fig. 5.3) (Purves et al., 2001).

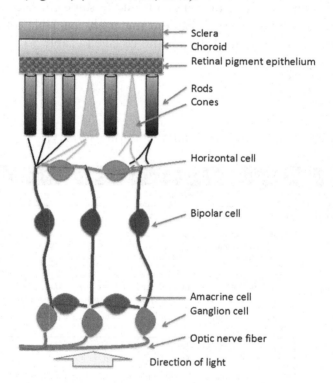

*Figure 5.3 Structure of the retina.*

## How many types of neurons are there in the retina?

Photoreceptors—rods and cones
  Bipolar cells
  Ganglion cells
  Horizontal cells
  Amacrine cells

## What are the two types of light-sensitive elements in the retina?

Rods: Photoreceptors specialized for operating at low light levels.

Cones: Photoreceptors specialized for high visual acuity and the perception of color.

Both types of photoreceptors have an outer segment composed of membranous disks (Fig. 5.4) that contain photopigment, and an inner segment that gives rise to synaptic terminals that contact bipolar or horizontal cells (Purves et al., 2001).

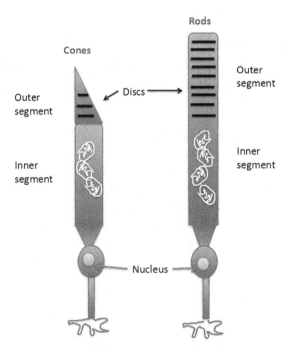

*Figure 5.4 Structure of the photoreceptors.*

## 5.5.1 Retinal Degeneration

Retinal degeneration can cause permanent visual loss, and affect millions worldwide. Common retinal diseases include age-related macular degeneration, the leading cause of visual impairment in the developed world; and retinitis pigmentosa, an inherited, degenerative disease that affects nearly 80,000 Americans. A feature common to this heterogeneous group of diseases is the irreversible loss of light-sensing photoreceptors (Trese et al., 2012).

**What are the two major strategies used to regenerate a degenerating retina?**

1. Photoreceptor transplantation (ie, transplant progenitor photoreceptor cells);
2. RPE transplantation. A defective RPE leads to photoreceptor degeneration. Transplantation of the RPE corrects this defect and prevents the photoreceptor cells from degenerating. Two different transplantation techniques are currently used in clinical studies: the RPE−cell suspension technique (cells delivered subretinally through small retinotomies) and the RPE−choroid−sheet technique (translocation of a cell sheet of RPE).

### Retinal Progenitor cells

Retinal progenitor cells (RPCs) are multipotent progenitor cells that can give rise to all the neurons of the retina.

## 5.5.2 Transplantation Strategies

**Retinal Sheet Transplantation**

Retinal sheet transplants are currently one of the few therapeutic options for progressive retinal degeneration. Retinal sheet transplants rely on the immature retinal sheet extending cell processes and forming synaptic connections with the degenerate host retina. Retinal sheet transplantation has the benefit of allowing the organized delivery of properly oriented cells to the areas of most severe retinal degeneration (Aramant and Seiler, 2002). However, in patients this technique has only shown subjective visual improvement (Humayun, 2000).

**Cell Transplantation**

A number of cellular sources that can be driven toward a fully differentiated RPE or photoreceptor cell fate have been studied. This includes embryonic stem cells, fetal tissue, progenitor cells, induced pluripotent stem cells, and adult tissue-specific stem cells. The donor cells must be at the correct stage in development at the time of transplantation (Pearson, 2014).

Retinal transplantations aimed at regenerating the sensory retina have largely been performed using neural and RPCs. Animal experiments over the last 30 years have demonstrated that RPCs can survive and differentiate after transplantation to the subretinal space, and this can lead to a degree of functional recovery (Klassen et al., 2004). However, RPCs delivered alone are often lost during the transplantation process. RPCs delivered on polymer scaffolds to dystrophic retina in mice have shown an increase in the percentage of RPC survival and provide a structure for high cell number delivery to regions of photoreceptor loss (Tao and Desai, 2007). Transplantation of cells into the subretinal space has been achieved safely in human subjects with high tolerance for graft tissue (Humayun et al., 2000). However, bolus injection of stem and progenitor cells to the subretinal space results in disorganized and poorly localized grafts, usually due to poor cell integration (Tomita et al., 2005). Techniques to reduce cell death include using injectable polymer cell delivery systems (Ballios et al., 2010).

**What are the ideal properties of polymers for retinal TE?**

- Biocompatibility in the eye, with little immune response in the subretinal space;
- Thinness ($<50$ μm thick);
- High porosity;
- Biodegradability (it must slowly disintegrate through hydrolysis);
- Young's modulus similar to the delicate sensory retina;
- Enough robustness to support surgical manipulation.

Trese et al. (2012).

**A number of polymers meet many of these requirements and have been approved by the U.S. Food and Drug Administration (FDA). List three polymers suitable for use in retinal TE and approved by the FDA?**

Poly(lactic-*co*-glycolic acid) (PLGA);
Poly(lactic acid) (PLLA);
Poly(caprolactone) (PCL).

When compared to bolus injection, the PLLA–PLGA polymer cell delivery system has been shown to increase cell survival and to increase the number of cells successfully delivered to the subretinal space (Tomita et al., 2005). A single polymer, however, is unlikely to meet the needs of both RPE and photoreceptor replacement therapy, and thus a combination of polymers with different properties may need to be used.

## 5.6 CONJUNCTIVAL TISSUE ENGINEERING

The conjunctiva consists of a conjunctival epithelium (nonkeratinized, stratified layer that together with the corneal epithelium provides stability to the tear film) and a vascularized stroma. The main functions of the conjunctiva are to secrete the mucin component of the tear film, form a protective barrier, and provide immune defense of the ocular surface (Schrader et al, 2009). Conjunctival damage can lead to damage to the ocular surface epithelia, erosion formation, ulcer formation, and bacterial infection (Ormerod et al, 1988). There is often a prolonged inflammation, which causes depletion of the limbal epithelial stem cell population and leads to corneal neovascularization, ingrowth of fibrous tissue, and stromal scarring (Shapiro et al, 1981). This can lead to ocular discomfort and blindness from corneal opacity.

Various techniques have been used to restore the corneal surface. These include autologous mucous tissue grafts such as conjunctival autografts for small defects, oral mucous membrane grafts for fornix reconstruction (Shore et al, 1992), nasal turbinate mucosal grafts (Kuckelkorn et al, 1996), which provide substitute mucin, and hard palate grafts for posterior lamella reconstruction (Ito et al, 2001). The latter has also been reconstructed using vein wall grafts (Barbera et al, 2008). All of the above techniques have limited applications due to the lack of suitable donor tissue when large grafts are required. Furthermore, autologous tissue may be impaired in patients suffering from autoimmune conditions. Tissue engineering of conjunctival substitutes aims to overcome this limitation. The tissue-engineered construct must meet several requirements. The material must be flexible, have good long-term stability, be biocompatible, and have an epithelial layer that has self-renewal potential. It should also contain epithelial cells and goblets cells, which are needed for the tear film and stability of the ocular surface (Schrader et al, 2009). Materials that may be useful include collagen matrix with epithelial cells and fibroblasts, and acellular polymers. The amniotic membrane (AM) is useful for conjunctival tissue engineering, particularly fornix reconstruction. It is an ideal conjunctival tissue substitute as it is thin, elastic, and does not cause rejection in the host. The AM is also a suitable carrier for the in vitro culture of conjunctival and oral mucosal epithelial cells. Although AM currently meets many of the criteria for an ideal conjunctival substitute, results in chronic inflammatory diseases have been less than satisfactory, and thus new materials must be developed to overcome this problem (Schrader et al., 2009).

## 5.7 SUMMARY

The World Health Organization reports a high prevalence worldwide of people who have impaired vision. Vision loss is a huge cause of morbidity for patients, and affects their livelihood, particularly those residing in developing countries. Corneal diseases are a major cause of vision loss and blindness. The applications of artificial corneas, along with studies of corneal TE, have brought hope to patients with corneal blindness. Retinal degeneration is also a leading cause of untreatable blindness. There are few effective treatments to replace lost photoreceptor cells and restore visual function. Transplanted donor photoreceptors have been shown to integrate with the neuronal circuitry of the recipient retina. Polymer delivery systems can be used to successfully deliver cells to the retina. It is anticipated that advances in TE of the various components of the eye will avoid the limitations of current treatments and provide new hope to many people worldwide.

## REFERENCES

Aramant, R.B., Seiler, M.J., 2002. Retinal transplantation—advantages of intact fetal sheets. Prog. Retin. Eye Res. 21, 57–73.

Avadhanam, V., Smith, H.E., Liu, C., 2015. Keratoprostheses for corneal blindness: a review of contemporary devices. Clin. Ophthalmol. 9, 697–720.

Ballios, B.G., Cooke, M.J., van der Kooy, D., Shoichet, M.S., 2010. A hydrogel-based stem cell delivery system to treat retinal degenerative diseases. Biomaterials 9, 2555–2564.

Barbera, C., Manzoni, R., Dodaro, L., Ferraro, M., Pella, P., 2008. Reconstruction of the tarsus-conjunctival layer using a venous wall graft. Ophthal. Plast. Recontr. Surg. 24 (5), 352–356.

Carrier, P., Deschambeault, A., Audet, C., Talbot, M., Gauvin, R., Giasson, C.J., et al., 2009. Impact of cell source on human cornea reconstructed by tissue engineering. Invest. Ophthalmol. Vis. Sci. 50 (6), 2645–2652.

Crawford, G.J., Hicks, C.R., Lou, X., Vijayasekaran, S., Tan, D., Mulholland, B., et al., 2002. The Chirila keratoprosthesis: phase I human clinical trial. Ophthalmology 109 (5), 883–889.

Ezhkova, E., Fuchs, E., 2010. An eye to treating blindness. Nature 466 (7306), 567–568.

Fagerholm, P., Lagali, N.S., Carlsson, D.J., Merrett, K., Griffith, M., 2009. Corneal regeneration following implantation of a biomimetic tissue-engineered substitute. Clin. Transl. Sci. 2 (2), 162–164.

Forrester, J.V., Dick, A.D., McMenamin, P.G., Roberts, F., Pearlman, E., 2016. The Eye: Basic Sciences in Practice. 4th Edition.

Gorovoy, M.S., 2006. Descemet-stripping automated endothelial keratoplasty. Cornea 25 (8), 886–889.

Humayun, M.S., de Juan Jr., E., del Cerro, M., Dagnelie, G., Radner, W., Sadda, S.R., et al., 2000. Human neural retinal transplantation. Invest. Ophthalmol. Vis. Sci. 41, 3100–3106.

Ito, O., Suzuki, S., Park, S., Kawazoe, T., Sato, M., Saso, Y., et al., 2001. Eyelid reconstruction using a hard palate mucoperiosteal graft combined with a V-Y subcutaneously pedicled flap. Br. J. Plast. Surg. 54 (2), 106–111.

Jay, L., Bourget, J.-M., Goyer, B., Singh, K., Brunette, I., Ozaki, T., 2015. Proulx. Characterization of tissue-engineered posterior corneas using second- and third-harmonic generation microscopy. PLoS One 10 (4), e0125564.

Klassen, H.J., Ng, T.F., Kurimoto, Y., Kirov, I., Shatos, M., Coffey, P., et al., 2004. Multipotent retinal progenitors express developmental markers, differentiate into retinal neurons, and preserve light-mediated behavior. Invest. Ophthalmol. Vis. Sci. 45 (11), 4167–4173.

Kuckelkorn, R., Schrage, N., Redbrake, C., Kottek, A., Reim, M., 1996. Autologous transplantation of nasal mucosa after severe chemical and thermal eye burns. Acta Ophthalmol. Scand. 74 (5), 442–448.

Ormerod, L.D., Fong, L.P., Foster, C.S., 1988. Corneal infection in mucosal scarring disorders and Sjögren's syndrome. Am. J. Ophthalmol. 105 (5), 512–518.

Pearson, R.A., 2014. Advances in repairing the degenerate retina by rod photoreceptor transplantation. Biotechnol. Adv. 32 (2), 485–491.

Purves, D., Augustine, G.J., Fitzpatrick, D., 2001. Neuroscience, second ed. Sinauer Associates, Sunderland, MA.

Schrader, S., Notara, M., Beaconsfield, M., Tuft, S.J., Daniels, J.T., Geerling, G., 2009. Tissue engineering for conjunctival reconstruction: Established methods and future outlooks. Current Eye Res. 34 (11), 913–924.

Shapiro, M.S., Friend, J., Thoft, R.A., 1981. Corneal re-epithelialization from the conjunctiva. Invest. Ophthalmol. Vis. Sci. 21 (1), 135–142.

Shore, J.W., Foster, C.S., Westfall, C.T., Rubin, P.A., 1992. Results of buccal mucosal grafting for patients with medically controlled ocular cicatricial pemphigoid. Ophthalmology 99 (3), 383–395.

Tao, S.L., Desai, T.A., 2007. Aligned arrays of biodegradable poly(epsilon-caprolactone) nanowires and nanofibers by template synthesis. Nano Lett. 7, 1463–1468.

Tomita, M., Lavik, E., Klassen, H., Zahir, T., Langer, R., Young, M.J., 2005. Biodegradable polymer composite grafts promote the survival and differentiation of retinal progenitor cells. Stem Cells 23, 1579–1588.

Trese, M., Regatieri, C.V., Young, M.J., 2012. Advances in retinal tissue engineering. Materials 5, 108–120.

Wang, H.Y., Wei, R.H., Zhao, S.Z., 2013. Evaluation of corneal cell growth on tissue engineering materials as artificial cornea scaffolds. Int. J. OPhthalmol. 6 (6), 873–878.

WHO, 2014. Visual impairment and blindness. Available online: <http://www.who.int/mediacentre/factsheets/fs282/en/>.

# CHAPTER 6

# Cardiovascular Tissue Engineering

## F. Akter[1] and H. Hamid[2]
[1]University of Cambridge, Cambridge, United Kingdom [2]Health Education East of England Deanery, Cambridge, United Kingdom

## 6.1 INTRODUCTION

Cardiovascular disease is one of the leading causes of death in the Western world (Moroni and Mirabella, 2014). Cardiovascular disease management has seen huge developments over the last 40 years with the advent of percutaneous coronary intervention and advancements in medical treatment. Death rates have fallen by more than two-thirds. However, there remains a substantial disease burden from chronic heart failure. Myocardial infarction (MI) results in scar formation over the ischemic myocardium and ventricular remodeling, which subsequently leads to ventricular dysfunction (Bolick et al., 1986). This manifests as heart failure and sudden cardiac death. There is limited capability for the adult heart to undergo endogenous cardiomyocyte proliferation and myocardial regeneration (Senyo et al., 2014).

There has been extensive research into stem cell therapy with the goal of inducing myocardial tissue repair and reversing the remodeling of the injured myocardium. There have been significant advances in mesenchymal stem cell (MSC) therapy with evidence of safe use, reversal of cardiac remodeling, and improvements in cardiac function. Research in this area has now moved into early clinical trials. The main limitation is stem cell survival posttransplantation. Further research is underway to try and achieve the holy grail of inducing complete myocardial regeneration without adverse consequences (Karikkineth and Zimmermann, 2013).

Cardiovascular tissue engineering (CTE) has emerged as an alternative cell-based approach, and is another exciting field in regenerative medicine. There are three main targets for CTE—bloods vessels, heart muscle, and heart valves—and the most commonly used biomaterials for engineering are biodegradable scaffolds, hydrogels, and decellularized tissue.

Tissue Engineering Made Easy. DOI: http://dx.doi.org/10.1016/B978-0-12-805361-4.00006-0

**Table 6.1 Cell Sources for Cardiovascular Tissue Engineering (Leor et al., 2005; Zhang et al., 2009; Galvez-Monton et al., 2013)**

| | Advantages | Disadvantages |
|---|---|---|
| Bone marrow-derived mesenchymal stem cells | Autologous Multipotent Low immune response | Cells can cause donor site morbidity during harvesting and may cause fibrosis |
| Adipose-derived mesenchymal stem cells | Easy to harvest | Low survival |
| Embryonic stem cells | Pluripotent Easy to expand | Teratogenic Ethical concerns |
| Induced pluripotent stem cells | Can be obtained in larger numbers for use in cell therapy and tissue engineering | Potentially teratogenic May be oncogenic |
| Skeletal myoblasts | Easily isolated | High incidence of arrhythmias |
| Fetal cardiomyoctes | Cardiomyocyte phenotype | Limited availability Ethical concerns |

## 6.2 MYOCARDIAL TISSUE ENGINEERING

### 6.2.1 Cell Sources

There a number of cell sources available for myocardial tissue engineering (Table 6.1). The optimal cell source to create for a myocardial construct must differentiate into cardiomyocytes, be easy to harvest, be nonimmunogenic, be able to expand in vitro on a large scale, and be able to integrate with the host tissue. The ideal donor cells are autologous cardiomyocytes; however, these are difficult to harvest. The most widely used stem cell source has been bone marrow. Bone marrow-derived MSCs show the ability to differentiate into multilineages, play a role in immunomodulation, and can secrete a host of growth factors that may induce endogenous myocardial regeneration (Williams et al., 2013).

#### 6.2.1.1 Induced Pluripotent Stem Cells

Induced pluripotent stem cells (iPSCs) can differentiate into cardiomyocytes (Yamakawa and Ieda, 2015). Cardiomyocyte therapy based on iPSCs has been limited to animal studies until there is better understanding of the behavior of transplanted iPSCs, further advances in isolation and purification techniques to reliably guide iPSCs to a cardiovascular lineage, and development of new techniques to make the iPSC-derived cardiomyocytes more mature and more similar to adult cardiac tissue (Csöbönyeiová et al., 2015).

### 6.2.1.2 Embryonic Stem cells

Embryonic stem cells (ESCs) are pluripotent cells, which can propagate without differentiation in cell culture and yet maintain the potential to differentiate into all three embryonic germ layers. ESCs transplanted into damaged heart result in differentiation of cardiomyoctes, endothelium, and vascular smooth muscle (Singla et al., 2006). ESCs transplanted into the infarcted heart can improve heart function (Hodgson et al., 2004). However, the amount of engraftment and differentiation into cardiomyoctes is limited. This may be improved by adding growth factors such as insulin-like growth factor-1 (IGF-1) to induce differentiation. Another limitation with ESCs is the danger of teratoma formation (Singla, 2009).

### 6.2.1.3 Mesenchymal Stem Cells

The cardioreparative effects of MSCs include their ability to stimulate angiogenesis, stimulate proliferation of endogenous cardiac stem cells, restore contractile function, and reduce fibrosis (Karantalis and Hare, 2015).

Mesenchymal stem cells and their subpopulations have been studied in numerous preclinical and early-phase clinical trials, providing a wealth of evidence on their mechanism of action, safety of use in transplantation, and potential therapeutic effects on cardiac regeneration. One of the main limitations reported with MSC therapy is the low survival rate of the transplanted cells and subsequent limitation on long-term efficacy. This has been addressed with more recent studies using pretreatment techniques with a cocktail of factors on the stem cells to enhance their retention rates and efficacy (Karantalis and Hare, 2015). The evidence so far suggests that complete myocardial regeneration cannot be achieved with only one cell type, therefore, future research may combine different cell types to enhance the therapeutic potential of stem cell therapy in cardiovascular regeneration (Karantalis and Hare, 2015).

### 6.2.1.4 Skeletal myoblasts

Skeletal myoblasts have been widely studied for use in myocardial repair. Skeletal myoblasts have a number of ideal properties for use in myocardial damage such as their autologous availability: they can develop into myofibers after engraftment into the heart; they are resistant to ischemia and have low risk of tumorigenesis (Durrani et al., 2010). However, a number of challenges limits their acceptance as ideal donor cells for myocardial repair. Myofibers derived from skeletal myoblasts cannot develop intercalated discs and consequently demonstrate poor electromechanical coupling with the host myocardium, which may contribute to arrhythmia development (Abraham and Hare, 2006). Another challenge with skeletal myoblasts is the high attrition rate following engraftment, which reduces the efficacy of the procedure (Suzuki et al., 2004).

### 6.2.1.5 Route of Cell Delivery

The efficacy of different stem cell delivery methods is dependent on the active disease process. In acute MI, the inflammatory cascade that leads to the release of chemokines, cytokines, and growth factors triggers the migration of endogenous resident stem cells as well as bone marrow-derived stem cells. In this setting, intracoronary or intravenous stem cell delivery methods are used (Karantalis and Hare, 2015). The limitations of the intravenous technique are the distribution of stem cells throughout the body and possible entrapment in other organs. The intracoronary route, on the other hand, requires a transient period of ischemia through balloon inflation to avoid the cells being washed out. Microvascular obstruction by infused cells may also lead to myocardial necrosis. In patients with advanced coronary disease, intracoronary delivery may not be feasible. In acute MI where there are areas of ischemia and necrosis, the intramyocardial delivery method is avoided due to concern about possible perforation (Karantalis and Hare, 2015). In chronic ischemic cardiomyopathy, the cues for stem cell migration are thought to be minimal, and therefore intramyocardial methods are preferred. Clinical trials have shown a correlation between injection site and improvement in contractility, and therefore better strategy design is recommended to achieve more precise delivery of the stem cell injections to cover the impaired myocardium (Karantalis and Hare, 2015).

### 6.2.2 Scaffolds

The aim of tissue engineering of the myocardium is to regenerate damaged cardiac tissue using tissue-engineered constructs, which mimic native extracellular matrix (ECM). Biomaterials need to be have mechanical properties to match those of native myocardium so that the delivered cells can integrate and remain intact in vivo. The constructs must be (1) contractile, (2) electrophysiologically stable, (3) robust yet flexible, (4) vascularized, and (5) nonimmunogenic. The construct

| Table 6.2 Strategies for Cardiac Tissue Engineering | |
| --- | --- |
| Strategies for Creating an Ideal Tissue-engineered Construct | |
| Scaffolds for cardiac tissue | Scaffolds to mimic the extracellular matrix (ECM) include natural materials—for example, alginate, collagen, or synthetic materials. The ideal artificial cardiac tissue must match the natural extracellular cardiac matrix, and the implanted cells must integrate fully (Galvez-Monton et al., 2013) |
| Decellularized tissues (ECM) | The ECM participates in many processes and cellular responses, including proliferation, differentiation, and migration. ECMs can be obtained from allogenic or xenogenic tissue to overcome the limitation of scarcity of autogenic tissue (Gilbert et al., 2006). These tissues must then be decellularized to eliminate allogenic and xenogenic antigens, and cell and nuclear content. However, they must preserve the properties and integrity of the ECM (Galvez-Monton et al., 2013) |
| Decellularized heart | Decellularization techniques can also be used for the whole heart. Rat cadaver hearts were decellularized to obtain an extracellular cardiac matrix with a preserved vascular tree, valves, and intact atrial and ventricular geometry. These constructs were then recellularized with endothelial cells and neonatal cardiac cells using coronary perfusion in a bioreactor to stimulate cardiac physiology. After 8 days of incubation and electrical stimulation, contractions and pump function were detected (Ott et al., 2008). Much remains to be done before a bioartificial heart is available for use in humans (Galvez-Monton et al., 2013) |
| Intramyocardial Injection of cells in hydrogel | Cells are embedded in natural hydrogel (eg, Matrigel™ (made of laminin, type IV collagen, and heparan sulfate), collagen, fibrin for intramyocardial injection) or synthetic hydrogels (polylactic acid-*co*-glycolic acid, polyethylene glycol, polylactic acid) or hybrid natural/synthetic hydrogels. Limitations of this technique include the need for high injection pressure, which can cause cell mortality (Galvez-Monton et al., 2013) |
| Ex vivo formation of hydrogel cell tissue | Tissues are created with cells that have been previously incorporated into hydrogel. However, whether these constructs can survive after mechanostimulation is yet to be seen (Galvez-Monton et al., 2013) |
| Monolayer cell constructs | Cells are cultured in polymer plaques followed by detachment of cell monolayers, which can then be implanted. The monolayers can generate intercellular communication that activates contractile function and propagates signals within the construct (Shimizu et al., 2002). However, whether this technique can be translated into the human heart is not yet known (Galvez-Monton et al., 2013) |

requires the use of a suitable cell source seeded onto a suitable scaffold; this has been demonstrated in animal studies where cardiomyocyte seeded within porous scaffolds yielded 3-D high-density cardiac constructs (Dar et al., 2002). A number of strategies for cardiac engineering are available (Table 6.2). Natural scaffolds such as collagen or synthetic polymers are usually seeded in vitro with cardiac cells to create contractile cardiac patches (Moroni and Mirabella, 2014). However, creating the ideal construct must overcome a number of challenges, such as adequate vascularization, adequate perfusion, cell survival, and integration of the engineered cardiac tissue (Leor et al., 2005). Furthermore, functional parameters such as heart rate, heart size, and regeneration capability of the construct must be fully determined in order to avoid complications such as arrhythmias (Jana et al., 2014).

## 6.2.3 Growth Factors

Various growth factors have been used to induce cardiomyocyte differentiation, such as vascular endothelial growth factor (VEGF), which is a potent mitogen for vascular endothelial cells (Chen et al., 2006); fibroblast growth factor (FGF) (Chan et al., 2010); the protein activin A; and bone morphogenic protein-4 (Suh et al., 2014).

## 6.2.4 Bioreactor

The bioreactor is a critical part in supporting growth of biological implants. Bioreactors combined with mechanical signals can improve the proliferation of seeded cells throughout the scaffold. Regulatory signals can be generated by secretion of growth factors and interstitial flow. Perfusion platforms to provide oxygen transport have been developed. Scaffolds were created with a preformed capillary-like channel to mimic native capillaries, and an oxygen carrier to perfuse the cell-seeded scaffolds and mimic the native oxygen carrier hemoglobin (Radisic et al., 2006).

Cardiogenesis is regulated by the interplay between biochemical, mechanical, and electrical stimuli. Mechanical stimulation is a critical aspect of cardiac function. Mechanical stretch of neonatal rat heart cells in collagen gels enabled development of constructs capable of generating substantial mechanical force (Zimmermann et al., 2000). Contraction of cardiomyocytes is elicited by binding of calcium to cardiac troponin C (cTnT). It may therefore be useful to create bioreactors with cTnT to enhance the contractile function of engineered cardiac muscle (Jana et al., 2014).

Excitation−contraction coupling (ECC), where the heart contracts in response to electrical cardiac signals, is essential for the pumping action of the heart, and thus constructs must be created to ensure electromechanical cellular coupling and adequate contractile function take place. In vitro, an electrical field can be applied within a bioreactor through the use of carbon rods connected to an external stimulator. Electrical stimulation upregulates cardiac genes, aligns cardiomyocytes, and electrically couples cells for enhanced contractility and function (Radisic et al., 2004). Monitoring of these processes using an implanted online monitoring system will be useful (Sanchez et al., 2012).

## 6.2.5 Vascularization

Vascularization is essential for creating functional cardiac tissue, as cardiomyoctes have a high metabolic rate. Engineered cardiac tissue is either vascularized prior to implantation or conditioned until being vascularized following implantation. For prevascularization, blood vessels could be created in the cardiac tissue prior to implantation. Delayed vascularization strategies include implanting engineered tissues containing growth factors to guide angiogenesis (Parsa et al., 2016).

Several methods of increasing vascularization of the constructs have been tested. These include incorporating growth factors such as VEGF or basic FGF in the scaffold to promote endothelial cell proliferation and vascular structure formation (Shen et al., 2008; Lai et al., 2006). Other techniques include the in vitro use of bioreactors to improve oxygen perfusion (Galvez-Monton et al., 2013).

## 6.3 VALVULAR TISSUE ENGINEERING

Valvular heart disease is a major health problem that results in substantial morbidity and death worldwide. Tissue engineering of valves is an alternative to current treatment modalities. Here cells are grown onto a biocompatible scaffold and proliferate and differentiate into a functional construct. The construct must be able to integrate into the host, remodel, and mimic the native heart valve. Two main types of scaffold have been developed: (1) acellular scaffolds and (2) artificial scaffolds (made from synthetic or natural polymers). Decellularized scaffolds are similar to the original valve structure and ECM molecules, and their mechanical stiffness is close to that of the native valve. Artificial scaffolds unfortunately cannot currently mimic the 3-D structure of the native valves (Jana et al., 2014).

## 6.4 BLOOD VESSEL ENGINEERING

Blood vessels are essential in the body and mediate gas exchange, nutrient and waste transport, and immune defense. They consist of endothelial cells, vascular smooth muscle cells in the middle layer, and fibroblasts and matrix in the outer layer (Nemeno-Guanzon et al., 2012).

Tissue-engineered blood vessels can be made using scaffold materials such as native matrix, synthetic polymers (eg, polyglycolic acid), and biological materials (eg, collagen). These vessels must be nonthrombogenic, nonimmunogenic, compatible at high blood flow rates, and must have a similar viscoelasticity to native vessels (Zhang et al., 2007).

Tissue constructs must be seeded with appropriate cells. The ideal cell source should be nonimmunogenic, functional, and easy to achieve and expand in culture. Nonimmunogenic autologous endothelial cells, and smooth muscle cells isolated from patients, are the ideal cell sources. These can be differentiated from stem cells such as iPSCs, embryonic stem cells, and MSCs (Zhang et al., 2007).

The first production of a completely biological tissue-engineered blood vessel, using cultured smooth muscle cells and endothelial cells in bovine collagen gels, was performed in 1986 (Weinberg and Bell, 1986). Since then there have been a number of studies that have demonstrated the successful creation of scaffolds useful for blood vessel engineering. However, scaffold-based constructs face a number of difficulties such as poor cell-to-cell interaction, poor alignment of ECM components, and a complex host response to scaffolds. This has led to the development of a number of scaffold-free techniques. The first scaffold-free tissue-engineered human blood vessel was created in 1998 (L'Heureux et al., 1998), and has since been replicated many times. Scaffold-free vessels can be created using techniques such as decellurization, self-assembly of ECM by cells, and stimulation via bioreactors (Norotte et al., 2009).

Decellularized vessels, which contain only natural ECM, have good biocompatibility. Decellularization is achieved by treating tissues with detergents, enzyme inhibitors, and buffers. Bioreactors help achieve a functional graft in culture. These vessel-reactors mimic the physiological vessel stimuli that a native vessel receives, such as cyclical strain and shear stress. Bioreactors can be used to remodel the ECM and help vessels mature (Zhang et al., 2007).

In addition to mechanical stimulation, chemical reagents and growth factors must also be added to the culture media. Growth factors such as transforming growth factor $\beta 1$ can increase ECM production and vasoreactivity. Additives such as retinoic acid and ascorbic acid have been shown to enhance mechanical properties of an engineered vessel graft (Zhang et al., 2007).

Engineered blood vessels must overcome a number of challenges before they can be used clinically. Adequate vascularization is required for longevity of scaffolds. Bioreactor designs must also be improved to effectively stimulate the cells in a cost-effective and timely manner. It is particularly important for engineered constructs to be used widely in acute injuries (Nemeno-Guanzon et al., 2012).

## 6.5 SUMMARY

After an ischemic insult to the heart, there is an inflammatory cascade that results in scar formation and adverse remodeling of the myocardium. The heart has little capacity to repair itself, resulting in loss of function. Extensive research has been undertaken into stem cell therapy with the goal of inducing cardiovascular repair. Several graft materials are currently used to replace diseased cardiac tissue and valvular segments. However, current implantable grafts can trigger an immune response. Furthermore, they cannot remodel, which is a problem, particularly in children. A fully functional tissue-engineered construct that truly recapitulates the native heart is currently not available. However, it is anticipated that with advances in vascularization, cell survival, and perfusion of the construct, this will be possible.

## REFERENCES

Abraham, M.R., Hare, J.M., 2006. Is skeletal myoblast transplantation proarrhythmic? The jury is still out. Heart Rhythm. 3 (4), 462–463.

Bolick, D.R., Hackel, D.B., Reimer, K.A., Ideker, R.E., 1986. Quantitative analysis of myocardial infarct structure in patients with ventricular tachycardia. Circulation 74, 1266–1279.

Chan, S.S., Li, H., Hsueh, Y.C., Lee, D.S., Chen, J.H., Hwang, S.M., et al., 2010. Fibroblast growth factor-10 promotes cardiomyocyte differentiation from embryonic and induced pluripotent stem cells. PLoS One 5 (12), e14414.

Chen, Y., Amende, I., Hampton, T.G., Yang, Y., Ke, Q., Min, J., et al., 2006. Vascular endothelial growth factor promotes cardiomyocyte differentiation of embryonic stem cells. Am. J. Physiol. 291 (4), H1653–H1658.

Csöbönyeiová, M., Polák, Š., Danišovič, L., 2015. Perspectives of induced pluripotent stem cells for cardiovascular system regeneration. Exp. Biol. Med. (Maywood) 240 (5), 549–556.

Dar, A., Shachar, M., Leor, J., Cohen, S., 2002. Optimization of cardiac cell seeding and distribution in 3D porous alginate scaffolds. Biotechnol. Bioeng. 80, 305–312.

Durrani, S., Konoplyannikov, M., Ashraf, M., Haider, K.H., 2010. Skeletal myoblasts for cardiac repair. Regen. med. 5 (6), 919–932.

Galvez-Monton, C., Prat-Vidal, C., Roura, S., Soler-Botija, C., Bayes-Genis, A., 2013. Update: innovation in cardiology (IV). Cardiac tissue engineering and the bioartificial heart. Rev. Esp. Cardiol. (Engl. Ed.) 66 (5), 391–399.

Gilbert, T.W., Sellaro, T.L., Badylak, S.F., 2006. Decellularization of tissues and organs. Biomaterials 27, 3675–3683.

Hodgson, D.M., Behfar, A., Zingman, L.V., Kane, G.C., Perez-Terzic, C., Alekseev, A.E., et al., 2004. Stable benefit of embryonic stem cell therapy in myocardial infarction. Am. J. Physiol. Heart Circ. Physiol. 287, H471–H479.

Jana, S., Tefft, B.J., Spoon, D.B., Simari, R.D., 2014. Scaffolds for tissue engineering of cardiac valves. Acta Biomater. 10 (7), 2877–2893.

Karantalis, V., Hare, J.M., 2015. Use of mesenchymal stem cells for therapy of cardiac disease. Circ. Res. 116 (8), 1413–1430.

Karikkineth, B.C., Zimmermann, W.H., 2013. Myocardial tissue engineering and heart muscle repair. Curr. Pharm. Biotechnol. 14 (1), 4–11.

Lai, P.H., Chang, Y., Chen, S.C., Wang, C.C., Liang, H.C., Chang, W.C., et al., 2006. Acellular biological tissues containing inherent glycosaminoglycans for loading basic fibroblast growth factor promote angiogenesis and tissue regeneration. Tissue Eng. 12, 2499–2508.

Leor, J., Amsalem, Y., Cohen, S., 2005. Cells, scaffolds, and molecules for myocardial tissue engineering. Pharmacol. Ther. 105 (2), 151–163.

L'Heureux, N., Pâquet, S., Labbé, R., Germain, L., Auger, F.A., 1998. A completely biological tissue-engineered human blood vessel. FASEB J. 12 (1), 47–56.

Moroni, F., Mirabella, T., 2014. Decellularized matrices for cardiovascular tissue engineering. Am. J. Stem Cells 3 (1), 1–20.

Nemeno-Guanzon, J.G., Lee, S., Berg, J.R., Jo, Y.H., Yeo, J.E., Nam, B.M., et al., 2012. Trends in tissue engineering for blood vessels. J. Biomed. Biotechnol. 2012, 956345.

Norotte, C., Marga, F.S., Niklason, L.E., Forgacs, G., 2009. Scaffold-free vascular tissue engineering using bioprinting. Biomaterials 30, 5910–5917.

Ott, H.C., Matthiesen, T.S., Goh, S.K., Black, L.D., Kren, S.M., Netoff, T.I., et al., 2008. Perfusion-decellularized matrix: using nature's platform to engineer a bioartificial heart. Nat. Med. 14, 213–221.

Parsa, H., Ronaldson, J., Vunjak-Novakovic, G., 2016. Bioengineering methods for myocardial regeneration. Adv. Drug. Deliv. Red. 96, 195–202.

Radisic, M., Park, H., Shing, H., Consi, T., Schoen, F., Langer, R., et al., 2004. Functional assembly of engineered myocardium by electrical stimulation of cardiac myocytes cultured on scaffolds. Proc. Natl. Acad. Sci. U.S.A. 101 (52), 18129–18134.

Radisic, M., Park, H., Chen, F., Salazar-Lazaaro, J.E., Wang, Y., Dennis, R., et al., 2006. Biomimetic approach to cardiac tissue engineering: oxygen carriers and channeled scaffolds. Tissue Eng. 12 (8), 2077–2091.

Sanchez, B., Guasch, A., Bogonez, P., Galvez, C., Puig, V., Prat, C., et al., 2012. Towards on line monitoring the evolution of the myocardium infarction scar with an implantable electrical impedance spectrum monitoring system. Conf. Proc. IEEE Eng. Med. Biol. Soc. 2012, 3223–3226.

Senyo, S.E., Lee, R.T., Kühn, B., 2014. Cardiac regeneration based on mechanisms of cardiomyocyte proliferation and differentiation. Stem Cell Res. 13 (3), 532–541.

Shen, Y.H., Shoichet, M.S., Radisic, M., 2008. Vascular endothelial growth factor immobilized in collagen scaffold promotes penetration and proliferation of endothelial cells. Acta Biomater. 4, 477–489.

Shimizu, T., Yamato, M., Isoi, Y., Akutsu, T., Setomaru, T., Abe, K., et al., 2002. Fabrication of pulsatile cardiac tissue grafts using a novel 3-dimensional cell sheet manipulation technique and temperature-responsive cell culture surfaces. Circ. Res. 90 (3), e40.

Singla, D.K., Hacker, T.A., Ma, L., Douglas, P.S., Sullivan, R., Lyons, G.E., et al., 2006. Transplantation of embryonic stem cells into the infarcted mouse heart: formation of multiple cell types. J. Mol. Cell Cardiol. 40, 195–200.

Singla, D.K., 2009. Embryonic Stem Cells in Cardiac Repair and Regeneration. Antioxid. Redox Signal. 11 (8), 1857–1863.

Suh, C.Y., Wang, Z., Bártulos, O., Qyang, Y., 2014. Advancements in induced pluripotent stem cell technology for cardiac regenerative medicine. J. Cardiovasc. Pharmacol. Ther. 19 (4), 330–339.

Suzuki, K., Murtuza, B., Beauchamp, J.R., Smolenski, R.T., Varela-Carver, A., Fukushima, S., et al., 2004. Dynamics and mediators of acute graft attrition after myoblast transplantation to the heart. FASEB J. 18 (10), 1153–1155.

Weinberg, C.B., Bell, E., 1986. A blood vessel model constructed from collagen and cultured vascular cells. Science 231 (4736), 397–400.

Williams, A.R., Hatzistergos, K.E., Addicott, B., McCall, F., Carvalho, D., Suncion, V., et al., 2013. Enhanced effect of combining human cardiac stem cells and bone marrow mesenchymal stem cells to reduce infarct size and to restore cardiac function after myocardial infarction. Circulation 127 (2), 213–223.

Yamakawa, H., Ieda, M., 2015. Strategies for heart regeneration: approaches ranging from induced pluripotent stem cells to direct cardiac reprogramming. Int. Heart J 56 (1), 1–5.

Zhang, W.J., Liu, W., Cui, L., Cao, Y., 2007. Tissue engineering of blood vessel. J. Cell. Mol. Med. 11 (5), 945–957.

Zhang, J., Wilson, G.F., Soerens, A.G., et al., 2009. Functional cardiomyocytes derived from human induced pluripotent stem cells. Circ. Res. 104 (4), e30–e41.

Zimmermann, W.H., Fink, C., Kralisch, D., Remmers, U., Weil, J., Eschenhagen, T., 2000. Three-dimensional engineered heart tissue from neonatal rat cardiac myocytes. Biotechnol. Bioeng. 68 (1), 106–114.

# CHAPTER 7

# Lung Tissue Engineering

**F. Akter[1], L. Berhan-Tewolde[2], and A. De Mel[3]**
[1]University of Cambridge, Cambridge, United Kingdom [2]St Mary's Hospital, London, United Kingdom [3]University College London, London, United Kingdom

## 7.1 INTRODUCTION

Lung disease is one of the most common causes of death in the United States. Common causes of lung disease include chronic obstructive pulmonary disease (COPD), which affects more than 11 million people in the United States (ALA, 2016). Contributing to the high mortality associated with lung disease is that lungs demonstrate minimal regenerative abilities. For patients with end-stage lung disease, lung transplantation is the only option; however, this is limited by a paucity of donor organs. The aim of lung tissue engineering is to provide an alternative solution to treat lung disease not suitable for conventional treatment (Calle et al., 2014). Lung engineering requires three essential components: a biocompatible scaffold, appropriate cells, and a bioreactor to cultivate the tissue.

## 7.2 ENDOGENOUS STEM AND PROGENITOR CELLS

The characterization of resident lung stem and progenitor cells is an important step toward the understanding of lung repair after injury or lung disease.

### 7.2.1 Proximal Airway

The proximal airway is made up of pseudostratified epithelium composed of a basal layer consisting of a discontinuous layer of basal cells and a luminal layer containing ciliated cells, Clara-like cells (secretory club cells), human goblet cells, and neuroendocrine cells that aggregate to form neuroendocrine bodies (Asselin-Labat and Filby, 2012) (Fig. 7.1).

Tissue Engineering Made Easy. DOI: http://dx.doi.org/10.1016/B978-0-12-805361-4.00007-2

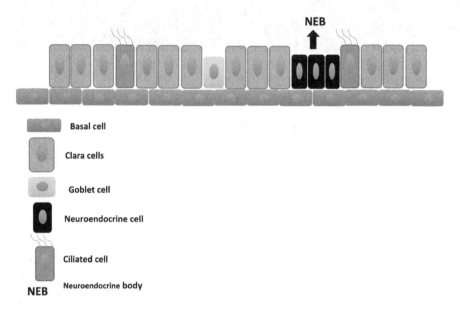

*Figure 7.1 Proximal airway pseudostratified epithelium composed of basal and luminal cells. Basal cells shown to act as stem cells during injury (Kotton and Morrisey, 2014).*

## 7.2.2 Distal Airway

The distal airway (small bronchi and bronchioles) consists of a single-layered epithelium containing Clara cells, ciliated cells, neuroendocrine cells, and $p63^+Krt5^+$ basal cells. Clara cells predominate over ciliated cells, and there are more neuroendocrine cells than in the trachea (Asselin-Labat and Filby, 2012) (Fig. 7.2).

## 7.2.3 Alveoli

Alveoli are present in the most distal region of the lung. Alveoli are composed of two types of epithelial cells: alveolar type I cells (AEC I), which provide a gas exchange surface; and cuboidal alveolar type II cells (AEC II), which containing secretory vesicles filled with surfactant (Asselin-Labat and Filby, 2012).

## 7.3 EXOGENOUS CELLS

### 7.3.1 Embryonic Stem Cells

Several studies have shown that embryonic stem cells (ESCs) can differentiate into progenitor and well-differentiated pulmonary epithelial cells (Huang et al., 2014; Akram et al., 2016). However, ESCs

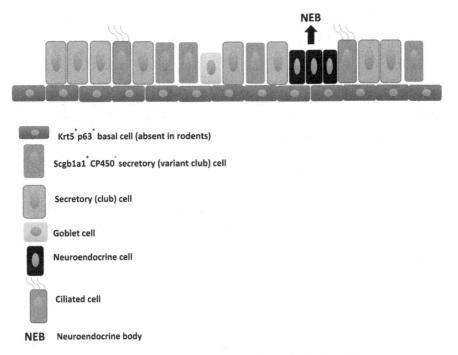

*Figure 7.2 Distal airway epithelium made up of luminal and basal cells (basal are Krt5⁺p63⁺ and absent in rodents). Variant club cells are Scgb1a1⁺ CP450⁻ secretory cells, which function as facultative progenitors after lung injury (Kotton and Morrisey, 2014).*

carry the risk of teratoma formation, and there are also ethical concerns regarding their use.

## 7.3.2 Induced Pluripotent Stem Cells

Induced pluripotent stem cells (iPSCs) are useful for researchers as they can provide an inexhaustible supply of cells that are identical to the patient's. In 2010, pluripotent reprogramming was used to derive the first 100 lung disease-specific iPSC lines from individuals with common conditions such as cystic fibrosis (Somers et al., 2010). One of the challenges that must be addressed is the characterization of in vitro iPSC-derived lung lineages, as most definitions of lung epithelia are based on their in vivo structure (Hawkins and Kotton, 2015).

## 7.3.3 Mesenchymal Stem Cells

Mesenchymal stem cells (MSCs) have been used in a number of disease models. Chronic lung conditions such as COPD, idiopathic pulmonary fibrosis (IPF), asthma, and pulmonary hypertension (PH) have been shown to benefit from treatment with MSCs.

COPD is a life-limiting lung disease where repeated cycles of broncho-alveolar destruction and repair lead to remodeling and eventual irreversible airway obstruction. MSCs have shown benefit in attenuating alveolar destruction in animal models. Different studies have shown this to be both via direct engraftment of the MSCs and differentiation into AEC II, and by attenuation of inflammation, inhibition of epithelial apoptosis, and by stimulation of alveolar and bronchiolar cell proliferation without direct engraftment. However, a recent placebo-controlled, randomized, multicenter clinical trial looking at treatment of COPD in humans with systemic MSCs showed no significant benefit or adverse effects (Akram et al., 2016; Zhen et al., 2008; Abreu et al., 2011; Yuhgetsu et al., 2006).

MSCs genetically engineered to overexpress endothelial nitric oxide synthase, prostacyclin, and heme oxygenase-1 have shown great benefit in the treatment of PH. Similarly, keratinocyte growth factors (KGFs) expressing MSCs given to mouse models with bleomycin-induced IPF have resulted in reduced fibrosis due to suppression of collagen accumulation. KGFs have also been shown to stimulate proliferation of AEC II cells (Ulich et al., 1994; Akram et al., 2016; Lee et al., 2012; Liang et al., 2011).

Acute respiratory distress syndrome (ARDS) is often a severe and life-threatening lung condition. It has several local and systemic causes, including infection, trauma, and surgery. Studies in animal models using MSCs to treat ARDS caused by endotoxin-induced lung injury, hyperoxia, pneumonia, and systemic sepsis have shown some benefits to MSCs (Akram et al., 2016).

The paracrine antiinflammatory properties of MSCs have been demonstrated in mouse models with lipopolysaccharide-induced acute lung injury (ALI). The mice were given intravenous or intratracheal MSCs, which significantly improved the ALI. This was likely via suppression of proinflammatory cytokine tumor necrosis factor-$\alpha$ and upregulation of antiinflammatory cytokine Interleukin-10. MSCs and acellular conditioned media from cultured MSCs have been found to attenuate acute pulmonary inflammation (Akram et al., 2016; Ortiz et al., 2007; Gupta et al., 2007). MSCs given to mice with allergic airway inflammation results in inhibition of the inflammation by inhibiting the T-helper cell (Goodwin et al., 2011).

Exogenous cell-based therapies have been shown to have some potential benefits in the treatment of both acute and chronic lung diseases. However, more studies need to be done, including optimal cell preparations, storage, dosing, and routes of administration, before they can be applied clinically (Weiss, 2014).

## 7.4 SCAFFOLDS

Scaffolds that can be created to improve lung regeneration must meet certain requirements:

- Must be biocompatible;
- Must remain in the tissue long enough to support cell growth but also be biodegradable;
- Must match lung properties (eg, possess sufficient elasticity to avoid developing restrictive lung disease);
- Must be porous to ensure adequate diffusion between air and blood;
- Must provide a suitable environment for cell adhesion and survival;
- Must be adapted to facilitate lung function and provide continuous ventilation;
- Must provide a large surface area for gaseous exchange and an intact basement membrane between air and blood;
- The vascular portion of the scaffold must contain patent, perfusable conduits so that blood can be oxygenated;
- Must be sterile with no toxic elements from manufacture retained within the scaffold.

    Calle et al. (2014)

Types of scaffolds used in lung regeneration include acellular scaffolds, and synthetic and natural polymers. Synthetic polymers used in lung tissue engineering include polyglycolic acid, and polylactic acids. However, they cannot match the composition of the natural matrix, and they do not support lung epithelial development when implanted in vivo. Natural materials that have been used to grow lung tissue include collagen, Matrigel, and compressed porcine skin gelatin (Gelfoam). However, these scaffolds have inadequate mechanical properties, variable degradation rates, and cannot fully recapitulate the complexity of the lung. They are, however, superior to synthetic polymer scaffolds in supporting lung epithelial development and vascularization. Due to its complex nature, the development of a scaffold formed from more than one material may be required to provide all the essential scaffolding requirements (Nichols et al., 2009; Calle et al., 2014).

### 7.4.1 Decellularized Lung Tissue

Several groups have demonstrated the efficacy of a decellularized extracellular matrix (ECM) as a scaffold for tissue engineering complex three-dimensional structures such as the lung. Niklason and colleagues were the first to demonstrate the success of acellular scaffolds. This group decellularized rat lungs using a detergent. A bioreactor was then used to culture pulmonary epithelium and vascular endothelium on the acellular lung matrix. The epithelium displayed a hierarchical branched organization within the matrix, and an intact basement membrane. The endothelial cells repopulated the vascular compartment. In vitro, the scaffolds of ECMs were similar to the native lung, and when implanted into rats in vivo, the engineered lungs participated in gas exchange for 45 minutes to 2 hours. However, after 3 hours, there was some evidence of clot formation in the lungs. This would need to be addressed before any translation into clinical trials (Petersen et al., 2010).

In 2008, the first example of the use of a decellularized cadaveric trachea to replace an airway segment in a patient was reported. The trachea was coated with chondrocytes derived from bone marrow-isolated MSCs. The coating was achieved by culturing the decellularized trachea with the chondrocytes seeded on the external surface, and airway epithelial cells seeded on the luminal side. The culture was kept in an air–liquid interface rotating bioreactor for 96 hours. This in vitro engineered trachea was then surgically inserted into the left main bronchus of the recipient (Macchiarini et al., 2008).

### 7.5 BIOREACTOR

Lung tissue engineering requires a bioreactor in which to cultivate the tissue. The bioreactor must provide oxygen and nutrients to the growing cells, remove waste, and provide mechanical stimuli such as an air–liquid interface. The bioreactor must be designed to provide adequate ventilation and perfusion of the cells (Calle et al., 2011). The frequency and volume of ventilation administered to tissues in vitro have significant impact on survival of the endothelial and epithelial layers (Birukova et al., 2008). Bioreactors, however, are costly, particularly due to the requirement for ventilators to mimic normal lung physiology. The use of bioactive factors (such as growth factors) for stem cell differentiation can also increase the cost. Current techniques to decellularize tissue causes substantial matrix damage due to the harsh conditions required to remove cellular material (eg, high pH, strong detergents), lengthy processing times, or preexisting tissue contamination from microbial colonization. (Birukova et al., 2008; Calle et al., 2011). Modifications of the decellularization technique have been described that maintain the global tissue architecture and still reduce tissue damage (Balestrini et al., 2015).

## 7.6 SUMMARY

The lungs are largely quiescent during adult life, except during times of injury. Most lung injuries are repaired with endogenous stem cells. However, there are diseases such as COPD, IPF, and PH that are life-limiting and are as yet lacking treatment that can reverse the damage. A great volume of work has been done looking into mechanisms of repairing and reversing severe lung injury, especially in animal models, using ESCs, iPSCs, and MSCs.

iPSCs have been generated from patients with acquired and inherited lung disease, and they have huge potential in stem cell therapy and tissue engineering due to the vast number of cells into which they can develop. The current challenge, however, is to generate fully functional epithelial, endothelial, and interstitial cells that compose the lungs. Tissue engineering strategies have been developed to increase regeneration in diseased lungs. A decellularized lung matrix bioreactor system has been created where a lung scaffold is seeded with cells and connected to a ventilation system. This system provides the potential for developing alternative treatments for chronic lung diseases that have not benefited from traditional treatment. The technique is currently still largely experimental, and will require significant modification before any clinical trials can commence.

# REFERENCES

Abreu, S.C., Antunes, M.A., Pelosi, P., Morales, M.M., Rocco, P.R., 2011. Mechanisms of cellular therapy in respiratory diseases. Intensive Care Med. 37, 1421–1431.

Akram, K.M., Patel, N., Spiteri, M.A., Rosyth, N.R., 2016. Lung regeneration: endogenous and exogenous stem cell mediated therapeutic approaches. Int. J. Mol. Sci. 17 (1), 128.

ALA, 2016. Lung health and diseases. American Lung Association. <http://www.lung.org/lung-health-and-diseases/>.

Asselin-Labat, M.-L., Filby, C.E., 2012. Adult lung stem cells and their contribution to lung tumourigenesis. Open Biol. 2 (8), 120094. Available from: http://dx.doi.org/10.1098/rsob.120094.

Balestrini, J.L., Gard, A.L., Liu, A., Leiby, K.L., Schwan, J., Kunkemoeller, B., et al., 2015. Production of decellularized porcine lung scaffolds for use in tissue engineering. Integr. Biol. 7 (12), 1598–1610.

Birukova, A.A., Rios, A., Birukov, K.G., 2008. Long-term cyclic stretch controls pulmonary endothelial permeability at translational and post-translational levels. Exp. Cell Res. 314, 3466–3477.

Calle, E.A., Petersen, T.H., Niklason, L.E., 2011. Procedure for lung engineering. J. Vis. Exp. JoVE. 8 (49), 2651.

Calle, E.A., Ghaedi, M., Sundaram, S., Sivarapatna, A., Tseng, M.K., Niklason, L.E., 2014. Strategies for whole lung tissue engineering. IEEE Trans. Bio-med. Eng. 61 (5), 1482–1496.

Goodwin, M., Sueblinvong, V., Eisenhauer, P., Ziats, N.P., Leclair, L., Poynter, M.E., et al., 2011. Bone marrow derived mesenchymal stromal cells inhibit th2-Mediated allergic airways inflammation in mice. Stem Cells 29, 1137–1148.

Gupta, N., Su, X., Popov, B., Lee, J.W., Serikov, V., Matthay, M.A., 2007. Intrapulmonary delivery of bone marrow-derived mesenchymal stem cells improves survival and attenuates endotoxin-induced acute lung injury in mice. J. Immunol. 179, 855–1863.

Hawkins, F., Kotton, D.N., 2015. Embryonic and induced pluripotent stem cells for lung regeneration. Ann. Am. Thorac. Soc. 12 (Suppl. 1), S50–S53.

Huang, S.X., Islam, M.N., O'Neill, J., Hu, Z., Yang, Y.G., Chen, Y.W., et al., 2014. Efficient generation of lung and airway epithelial cells from human pluripotent stem cells. Nat. Biotechnol. 32, 84–91.

Kanki-Horimoto, S., Horimoto, H., Mieno, S., Kishida, K., Watanabe, F., Furuya, E., et al., 2006. Implantation of mesenchymal stem cells overexpressing endothelial nitric oxide synthase improves right ventricular impairments caused by pulmonary hypertension. Circulation 114, I181–I185.

Kotton, D.N., Morrisey, E.E., 2014. Lung regeneration: mechanisms, applications and emerging stem cell populations. Nat. Med. 20 (8), 822–832.

Lee, C., Mitsialis, S., Aslam, M., Vitali, S., Vergadi, E., Konstantinou, G., et al., 2012. Exosomes mediate the cytoprotective action of mesenchymal stromal cells on hypoxia-induced pulmonary hypertension. Circulation 126, 2601–2611.

Liang, O., Mitsialis, S., Chang, M., Vergadi, E., Lee, C., Aslam, M., et al., 2011. Mesenchymal stromal cells expressing heme oxygenase-1 reverse pulmonary hypertension. Stem Cells 29, 99–107.

Macchiarini, P., Jungebluth, P., Go, T., Asnaghi, M.A., Rees, L.E., Cogan, T.A., et al., 2008. Clinical transplantation of a tissue-engineered airway. Lancet 372 (9655), 2023–2030.

Nichols, J.E., Niles, J.A., Cortiella, J., 2009. Design and development of tissue engineered lung: progress and challenges. Organogenesis 5 (2), 57–61.

Ortiz, L.A., Dutreil, M., Fattman, C., Pandey, A.C., Torres, G., Go, K., et al., 2007. Interleukin 1 receptor antagonist mediates the anti-inflammatory and anti-fibrotic effect of mesenchymal stem cells during lung injury. Proc. Natl. Acad. Sci. U.S.A. 104, 11002–11007.

Petersen, T.H., Calle, E.A., Zhao, L., et al., 2010. Tissue-engineered lungs for in vivo implantation. Science (New York, NY) 329 (5991), 538–541.

Somers, A., Jean, J.C., Sommer, C.A., Omari, A., Ford, C.C., Mills, J.A., et al., 2010. Generation of transgene-free lung disease-specific human iPS cells using a single excisable lentiviral stem cell cassette. Stem Cells 28, 1728–1740.

Ulich, T.R., Yi, E.S., Longmuir, K., Yin, S., Biltz, R., Morris, C.F., et al., 1994. Keratinocyte growth factor is a growth factor for type II pneumocytes in vivo. J. Clin. Invest. 93 (3), 1298–1306.

Weiss, D.J., 2014. Current status of stem cells and regenerative medicine in lung biology and diseases. Stem Cells 32 (1), 16–25.

Yuhgetsu, H., Ohno, Y., Funaguchi, N., Asai, T., Sawada, M., Takemura, G., et al., 2006. Beneficial effects of autologous bone marrow mononuclear cell transplantation against elastase-induced emphysema in rabbits. Exp. Lung Res. 32, 413–426.

Zhen, G., Liu, H., Gu, N., Zhang, H., Xu, Y., Zhang, Z., 2008. Mesenchymal stem cells transplantation protects against rat pulmonary emphysema. Front. Biosci. 13, 3415–3422.

# CHAPTER 8

# Bone and Cartilage Tissue Engineering

**F. Akter[1] and J. Ibanez[2]**
[1]University of Cambridge, Cambridge, United Kingdom [2]Guys and St Thomas Hospital, London, United Kingdom

## 8.1 INTRODUCTION

Bone tissue engineering (BTE) is an emerging field that aims to combat the limitations of conventional treatments of bone disease. Bone is a vascularized tissue that must provide a firm structural support, withstand load bearing, and rapidly respond to metabolic demand (Amini et al., 2012).

Bone defects are one of the leading causes of disability in elderly patients, leading to a decrease in quality of life. There is an increased need for bone replacement in bone diseases, osteoporosis-related fractures, trauma, congenital bone malformations, and tumor resections (Wang et al., 2014). Worldwide, an estimated 2.2 million bone graft procedures are performed annually. Autografts are the gold standard for treatment of bone defects, but there is a limited supply, and donor site morbidity is a significant problem. Bone allografts are alternatives, but they are expensive, and there is a risk of disease transmission and adverse host immune response (Fu et al., 2011).

Tissue engineering of the bone aims to induce new functional tissue that has integrated with the host without causing any reaction. The normal anatomy and functions of the skeleton are reviewed first, followed by a discussion of the principles of BTE.

## 8.2 BONE ANATOMY

There are a total of 213 bones in the human body. Bone is categorized into four types (Table 8.1) (Clarke, 2008). It has an organic−inorganic nanocomposite structure. The organic structure is mainly composed of type I collagen, and the inorganic phase consists of hydroxyapatite

Tissue Engineering Made Easy. DOI: http://dx.doi.org/10.1016/B978-0-12-805361-4.00008-4

**Table 8.1 Categories of Bone**

| Categories of Bone | Examples |
|---|---|
| Long bone | Upper body<br>Humeri, clavicles, radii, ulnae, and metacarpals |
| | Lower body<br>Femurs, tibiae, fibulae, metatarsals, and phalanges |
| Short bone | Carpal and tarsal bones, patellae, and sesamoid bones |
| Flat bone | Skull, mandible, scapulae, sternum, and ribs |
| Irregular bone | Vertebrae, sacrum, coccyx, and hyoid bone |

(HA) crystals, which are embedded between the collagen fibers (Wang et al., 2014).

## 8.2.1 Structure

The adult human skeleton is composed of 80% cortical bone (also known as compact bone) and 20% trabecular bone (also known as cancellous or spongy bone). Cortical bone forms the outer shell of most bones, known as the cortex. It is much denser and stronger than trabecular bone. Trabecular bone is a much softer bone in comparison to cortical bone. It makes up the inner layer of the bone and has a spongy, honeycomb-like structure (Clarke, 2008).

The primary functional unit of cortical bone is the osteon (Fig. 8.1). The osteon is a cylindrical structure that consists of concentric layers known as lamellae surrounding a central canal, the haversian canal (which contains the blood supply of the bone). Situated between the lamellae are the lacunae, which contain the osteocytes. The lacunae are connected to each other via canaliculi. Osteocyte filopodia, radiating processes of the osteocytes, project into these canaliculi. Trabeculae are made up of semilunar subunits, also known as cancellous osteons (Mundy and Martin, 1993).

**How are cortical bone and trabecular bone normally formed?**

They are formed in a lamellar pattern in which collagen fibrils are laid down in alternating orientations, which leads to significant strength. This normal lamellar pattern is, however, absent in woven bone, in which the collagen fibrils are laid down in a haphazard manner. Woven bone is therefore weaker than lamellar bone. Woven bone is usually formed during formation of primary bone, and also in diseased

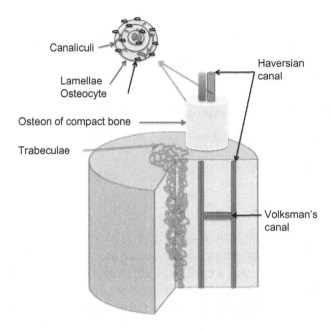

Canaliculi
Lamellae
Osteocyte
Osteon of compact bone
Trabeculae
Haversian canal
Volksman's canal

*Figure 8.1 Structure of cortical bone.*

| Table 8.2 Differences Between Periosteum and Endosteum (Netter, 1987) | |
|---|---|
| **Periosteum** | **Endosteum** |
| Fibrous connective tissue sheath | Membranous structure |
| Surrounds outer cortical bone surface | Covers the inner surface of cortical bone, trabecular bone, and the blood vessel canals (Volkmann's canals) |
| Attached to outer cortex by Sharpey's fibers (collagenous fibers) | Contains blood vessels, osteoblasts, and osteoclasts |

states with high bone turnover, such as osteitis fibrosa cystica and Paget's disease (Eriksen et al., 1994).

**Cortical bone has an outer periosteal surface and inner endosteal surface. What is the difference between periosteum and endosteum?**

Refer Table 8.2.

## 8.3 OSTEOGENESIS (BONE FORMATION)

Bone formation occurs via two pathways, known as intramembranous and endochondral ossification (Gilbert and Sunderland, 2000).

Intramembranous bone formation involves mesenchymal progenitor cells differentiating directly into osteoblasts, followed by development of the flat bones of the skull, mandible, maxilla, and clavicles (Gilbert and Sunderland, 2000).

The steps in intramembranous ossification are as follows:

1. Ossification center formation.
2. Calcification.
3. Trabeculae formation.
4. Periosteum development.

Endochondral bone formation (Fig. 8.2) involves mesenchymal progenitor cells differentiating into chondrocytes. Chondrocytes are responsible for depositing a cartilaginous template that is later mineralized and replaced by bone. This type of bone formation occurs in long bones (Gilbert and Sunderland, 2000). Several growth factors—such as bone morphogenetic proteins (BMPs), vascular endothelial growth factor (VEGF), and fibroblast growth factors (FGFs)—are important regulators in both processes (Gilbert and Sunderland, 2000).

If the process of formation of bone tissue occurs at an extraskeletal location, it is called heterotopic ossification.

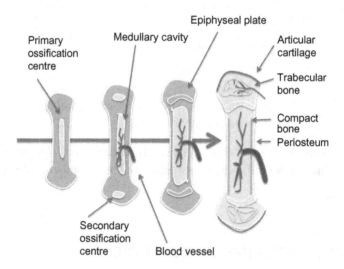

*Figure 8.2 Endochondral bone formation.*

Steps in osteogenesis:

1. Mesenchymal stem cells (MSCs) differentiate into osteoblasts.
2. Osteoblasts deposit extracellular organic matrix (osteoid).
3. Osteoblasts become trapped within the bone matrix and become osteocytes.
4. Matrix mineralization leads to the formation of bone.
5. Remodeling of bone occurs by the processes of resorption and reformation.

## 8.4 BONE TISSUE ENGINEERING

As with all forms of tissue engineering, the essential components required for bone formation include appropriate cells, a biocompatible scaffold, growth factors, and vascularization to meet the growing tissue nutrient supply.

**What are osteogenic cells?**

Osteogenic cells (osteoprogenitor cells) are bone cells that originate from MSCs. They can multiple continuously and differentiate into osteoblasts, which are responsible for forming bone. They are located in the periosteum, endosteum, and the stromal component of bone marrow (Clarke, 2008).

Bone consists of four types of cells: osteoblasts, osteoclasts, osteocytes, and osteoprogenitor (or osteogenic) cells (Table 8.3).

| Table 8.3 Bone Cells (Clarke, 2008; Kahn and Simmons, 1975; Boyle et al., 2003) | |
|---|---|
| **Types of Cells** | |
| Osteogenic cells | Cells that can multiple continuously and differentiate into osteoblasts |
| Osteoblasts | Synthesize proteins such as collagen, osteocalcin, and osteopontin, which compose the organic matrix of bone. The differentiation of mature osteoblasts is regulated by various pathways such as the Wnt pathway |
| Osteoclasts | Large multinucleated cells;Located in Howship lacunae;Produce enzymes (eg, acid phosphatase) that dissolve components of the bone |
| Osteocytes | Mature osteoblasts that have become trapped within the bone matrix they produced |

**What is osteoconduction?**

Osteoconduction refers to the ingrowth of capillaries and cells to support new bone formation on the implanted material. In BTE, following implantation of the material into the host, a fibrin clot forms. Osteogenic cells migrate to the surface of the scaffold through the fibrin clot (Amini et al., 2012).

**What is osteoinduction?**

Osteoinduction is the process by which MSCs are induced to differentiate into chondroblasts or osteoblasts. Growth factors such as BMPs can influence the recruitment and differentiation of MSCs (Amini et al., 2012).

## 8.4.1 Cells for BTE

In choosing an appropriate cell source for bone engineering strategies, one must consider the capacity of the chosen cells to differentiate into cells that can produce bone. A number of cell types have been shown to promote bone regeneration. These include MSCs, embryonic stem cells (ESCs), induced pluripotent stem cells (iPSCs) and adipose-derived stem cells (ADSCs).

There are various mechanisms by which implanted cells enhance bone regeneration in BTE. These involve release of key osteogenic and growth factors, recruitment of host osteogenic cells, and laying down bone matrix (Amini et al., 2012).

### 8.4.1.1 Mesenchymal Stem Cells

MSCs are multipotent cells that can be isolated from the bone marrow. The advantages of MSCs are their differentiation potential and their ability to expand extensively in vitro. They may also have immunosuppressive roles in vivo (Hwang et al., 2009).

**What are some of the difficulties in using MSCs for bone regeneration?**

- MSCs do not readily migrate to the bone.
- The bone tissue produced using current strategies is not adequately stiff to serve load-bearing functions in large defects in the body.

## How do MSCs enhance bone formation?

MSCs allow increased osteoinductivity of the biomaterial via the release of osteogenic growth factors and stimulation of the migration and differentiation of host osteoprogenitors (Vo et al., 2012).

## How do MSCs differentiate in vitro?

MSCs differentiate in three stages (Huang et al., 2007).

| Stage one | Days 1–4 |
|---|---|
| | Peak number of cells is seen |
| Stage two | Days 5–14 |
| | Early differentiation |
| | Characterized by the transcription and protein expression of alkaline phosphatase (ALP) (Aubin, 2001) |
| | Also found at an early stage is the expression of a collagen type I matrix onto which the mineral is deposited (Quarles et al., 1992) |
| Stage three | Days 14–28 |
| | High expression of osteocalcin and osteopontin, followed by calcium and phosphate deposition (Hoemann et al., 2009; Huang et al., 2007) |

## What does the differentiation of MSCs in vitro depend on?

The culture conditions:

1. MSCs + dexamethasone, ascorbic acid, and $\beta$-glycerol phosphate $\rightarrow$ osteogenic differentiation (Jaiswal et al., 1997).
2. MSCs + growth factors (eg, transforming growth factor-$\beta$ (TGF-$\beta$) family $\rightarrow$ chondrogenic differentiation) (Mackay et al., 1998).
3. MSCs + dexamethasone, insulin, isobutyl methyl xanthine, and indomethacin $\rightarrow$ adipogenic differentiation (Pittenger et al., 1999).

Typically, in vitro osteogenic factors are introduced directly into the culture medium of MSCs to drive the MSCs toward osteogenic differentiation. Recent in vivo studies have shown that culturing MSCs in the presence of bone cells can direct MSCs down the osteoblast lineage in the absence of osteogenic media, which could be useful for tissue engineering applications (Birmingham et al., 2012).

### 8.4.1.2 Embryonic Stem Cells

ESCs are pluripotent stem cells derived from the inner cell mass of a blastocyst.

**Describe some issues that limit the use of ESCs.**

Ethical issues regarding the use of embryos.

Risk of immune rejection after transplantation.

**How is osteogenic differentiation of ESCs achieved?**

Osteogenic differentiation of ESCs can be achieved by introducing cells in an osteoinductive medium.

ESCs can be directed to differentiate into the osteogenesis pathway by including factors in the culture medium, such as $\beta$-glycerophosphate, simvastatin, and BMPs (Bielby et al., 2004; Woll and Bronson, 2006; Pagkalos et al., 2010; Lee et al., 2013). Two methods have been examined in an attempt to stimulate the differentiation of human ESCs into osteogenic cells.

When cultured as single cells, ESCs spontaneously aggregate to form spheroids called embryoid bodies. In one method, osteogenic cells are derived directly from embryoid bodies. ESCs within embryoid bodies then undergo differentiation along all three germ lineages (endoderm, ectoderm, and mesoderm). In another method, ESC colonies are immediately separated into single cells, and then plated directly into a cell adhesive culture dish. This method has been tested but not widely used. It has the advantage of bypassing embryoid bodies and instead directing ESCs to differentiate into osteogenic cells immediately (Shimko et al., 2004; Sottile et al., 2003).

### 8.4.1.3 Adipose-derived Stem Cells

ADSCs represent an easily accessible and abundant source of autologous osteogenic cells. They have multilineage differentiation potential (ie, osteogenic, chondrogenic, adipogenic, neural, cardiomyocyte, and endothelial lineages) (Meyer et al., 2009). They have been demonstrated to undergo osteogenesis rapidly, and thus are a promising option for BTE trials (Cowan et al., 2004).

**How can ADSCs be isolated?**

ADSCs can be isolated by harvesting adipose tissue as a lipoaspirate. The specimen is microdissected, followed by serial washings and the use of collagenase to digest the stromal-vascular component, where ADSCs lie. The lipoaspirates are then centrifuged and the cell population is expanded in culture (Meyer et al., 2009).

**What mechanisms drive the ADSCs into the osteoblast lineage?**

Multiple signaling pathways have been demonstrated to participate in the differentiation of an osteoblast progenitor into a committed osteo-blast, including TGF-β/BMPs, Wnt/β-catenin, Notch, Hedgehog, and FGFs (Grottkau and Lin, 2013).

### 8.4.1.4 Induced Pluripotent Stem Cells

iPSCs have been successfully differentiated into osteoblast-like cells by using adenovirus vectors as gene delivery vehicles (Tashiro et al., 2009). Once differentiated into MSCs and subsequently terminally dif-ferentiated into functional osteoblasts, iPSCs have been shown to maintain their phenotype on a three-dimensional gelatin scaffold in vitro and in vivo (Bilousova et al., 2011).

## 8.4.2 Growth Factors for BTE

Bone contains numerous growth factors, which can function as signaling molecules. The most common growth factors that have been used for BTE include BMPs, TGF-β, FGFs, insulin-like growth factors I and II (IGF I/II), and platelet-derived growth factor (PDGF) (Table 8.4).

## 8.4.3 Scaffolds in BTE

Scaffolds have been used in repairing various clinical conditions in orthopedic surgery, maxillofacial surgery, and dentistry. Scaffolds play an important role in successful integration of the bone-grafting mate-rial in the host tissue, and provide mechanical support to the skeleton (Amini et al., 2012). Scaffolds should be rigid and resilient as they are the main supporting framework of bone grafts. Scaffolds must be porous with good pore interconnectivity for cell migration and forma-tion of 3D bone tissue in vitro and in vivo (Roosa et al., 2010). The porosity of a scaffold can also enhance osteoblast proliferation, and thus the extent of osteogenesis during bone regeneration (Kasten et al., 2008). Several scaffold materials have been used for BTE applications. Materials used for BTE scaffolds include the following:

Scaffolds must be...

1. Osteoinductive (capable of promoting the differentiation of progenitor cells down an osteoblastic lineage);
2. Osteoconductive (support bone growth and encourage the ingrowth of surrounding bone);
3. Capable of osseointegration (integration into surrounding bone).

1. *Natural or synthetic polymers such as hydrogels*

Polymers applied to BTE include biological polymers (such as collagen and hyaluronic acid), and synthetic polymers such as polylactic acid (PLA), polyglycolic acid (PGA), and polylactic-*co*-glycolic acid (PLGA) (Dhandayuthapani et al., 2011). Hydrogels are hydrophilic polymers that can be formed using natural materials, synthetic materials, or some combination of the two. Hydrogels are recommended for bone regeneration due to their resemblance to extracellular matrix

| Table 8.4 Growth Factors Used in BTE (Makhdom and Hamdy, 2012; Miyazono et al., 2010; Amini et al., 2012; Zhang et al., 2002; Govoni et al., 2007; Nakamura et al., 1996; Kawaguchi et al., 2010; Dai and Rabie, 2007; Ochman et al., 2011; Hollinger et al., 2008; Seifert et al., 1989) | |
|---|---|
| **Growth Factors** | |
| Bone morphogenetic proteins (BMPs) | • Proteins from the TGF-$\beta$ super-family.<br>• BMP signals (mediated by BMP receptors (BMPRs)) are transmitted by Smad proteins.<br>• Responsible for matrix formation and osteoblast formation.<br>• Express osteogenic markers such as ALP and osteocalcin through the mitogen-activated protein kinase pathway.<br>• BMPs 2, 4, 6, and 7 are considered to be the most osteoconductive of the BMPs.<br>• BMP-2, specifically, promotes undifferentiated mesenchymal cells into osteoblasts, leading to bone formation. |
| Transforming growth factor beta (TGF-$\beta$) | • TGF-$\beta$s are 25-kDa proteins.<br>• Three isoforms have been found (TGF-$\beta$1, $\beta$2, and $\beta$3).<br>• TGF-$\beta$ relevance to BTE is demonstrated by its ability to stimulate osteoblast-like cells to proliferate, and promotes collagen production in vitro.<br>• Play an integral part in collagen production and upregulate the expression of noncollagenous extracellular matrix proteins implicated in the regulation of bone turnover and mineralization. |
| Insulin-like growth factor (IGF) | • IGF is a chemotactic factor for osteoblasts.<br>• Osteoblast-specific knockout mice have also shown a decrease in cancellous bone volume.<br>• Conditional deletion of the IGF-1 in mice decreases bone formation. |
| Fibroblast growth factor (FGF) | • Compromise 22 members.<br>• Binds to FGF receptors.<br>• Locally applied FGF injection in osteoporotic rabbits enhanced bone formation.<br>• Local application of FGF hydrogel in a human trial enhanced fracture healing. |
| Vascular endothelial growth factor (VEGF) | • VEGF-related molecules include seven mammalian members: VEGF-A, -B, -C, -D, -E, and -F; and placenta growth factor.<br>• VEGF receptors include VEGFR-1/Flt-1, VEGFR-2/Flk-1, VEGFR-3/Flt-4, and two co-receptors of neuropilin and heparan sulfate proteoglycans.<br>• VEGF-A is most predominately involved in bone regeneration.<br>• Local application of VEGF in bone segment defects in rabbits has shown enhanced bone regeneration. |

*(Continued)*

| Table 8.4 (Continued) | |
|---|---|
| **Growth Factors** | |
| Platelet-derived growth factor (PDGF) | • The family of PDGFs includes PDGF-A, B, C, and D.<br>• The PDGFs signal through two cell-surface receptors: PDGF-R alpha and PDGF-R beta.<br>• PDGF is a potent chemoattractant for various mesenchymal cells, including osteogenic cells.<br>• PDGF can modulate the bone regenerative process via other growth factors (eg, by increasing VEGF signaling).<br>• Local PDGF implantation of demineralized bone matrix induces the local formation of cartilage and bone. |

(ECM), capacity to deliver bioactive molecules, and eventual biodegradability. Mechanical properties of hydrogels can be improved by combining the hydrogels with ceramic materials, such as calcium phosphate, β-tricalcium phosphate, and HA (Ahmed, 2015; Park, 2011).

2. *Bioactive ceramics such as bioactive glasses*
    Bioglass material is composed of minerals that occur naturally in the body ((silicon dioxide ($SiO_2$), calcium (Ca), sodium oxide ($Na_2O$), phosphorous (P))); the molecular proportions of the calcium and phosphorous oxides are similar to those in the bones. They bond strongly with bone by developing bone-like apatite layers on their surface in vivo. Bioactive glasses also release ions that activate expression of osteogenic genes and stimulate angiogenesis. The application of glass scaffolds for the repair of load-bearing bone defects, however, is often limited by their low mechanical strength and brittleness (Fu et al., 2011; Wu et al., 2012; Bose et al., 2012).
    Ceramic scaffolds, such as HA and tricalcium phosphate, are used for bone regeneration applications. Various ceramics have been used in dental and orthopedic surgery to fill bone defects and to coat metallic implant surfaces to improve implant integration with the host bone.

**What are the mechanical properties of ceramic scaffolds?**
- High mechanical stiffness (*Young's modulus*).
- Low elasticity.
- Hard brittle surface.
- Excellent biocompatibility in bones due to their chemical and structural similarity to the mineral phase of native bone.
- Enhancement of osteoblast differentiation and proliferation.

**Why are the clinical applications of ceramics for tissue engineering limited?**
- Brittleness.
- Difficult to shape for implantation.

- New bone formed in a porous HA network cannot sustain the mechanical loading needed for remodeling.
- Difficult to control degradation rate of HA.

**Definition: Young's Modulus**

*Young's modulus (elastic modulus)* is defined as the relationship between stress (force per unit area) and strain (proportional deformation) in a material.

A stiff material needs more force to deform compared to a soft material. Therefore, the Young's modulus is a measure of the stiffness of a solid material.

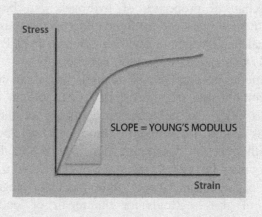

3. *Composites of polymers and ceramics*

   Composites are those that are made of two or more distinctly different materials such as ceramics and polymers to enhance mechanical properties (Bose et al., 2012).

4. *Metallic materials such as titanium alloys*

   There are various metallic scaffolds that have been used as implants. These include stainless steel, cobalt-based alloys, and titanium alloys. One of the main disadvantages of metallic biomaterials is their lack of biological recognition on the material surface. This can be overcome by surface coating or surface modification which, can be used to improve surface biocompatibility. Another limitation of metallic biomaterials is the possible release of toxic metallic ions through wear, which reduces their biocompatibility and causes tissue loss (Das et al., 2008; Alvarez and Nakajima, 2009).

## 8.4.4 Vascularization

Adequate vascularization within bone-engineered constructs is important to enable adequate implant survival and integration. In vivo conditions, oxygen, and nutrient supply are essential for the survival of growing cells and tissues within scaffolds. Poor angiogenesis has been identified as the main reason for implant failure, and is currently acknowledged as a major challenge in tissue engineering. The scaffolds must be designed to allow vascular ingrowth through a macroporous structure with interconnectivity of pores to allow cell migration. Incorporation of angiogenic factors into the scaffold enhances vascularization. Factors that induce angiogenesis in bone include VEGF and PDGF. The dose of the factor must optimized, as VEGF overexpression can lead to development of malformed and leaky vessels. In vitro prevascularization strategies include seeding and coculture of endothelial cells and osteogenic cells in the BTE constructs (Amini et al., 2012; Nguyen et al., 2012; Mastrogiacomo et al., 2006).

## 8.5 TENDON TISSUE ENGINEERING

Tendon injuries can be debilitating. Common tendon injuries include the flexor and extensor tendons of the fingers. Autografting is commonly required to repair an injury to a flexor tendon. However, limitations include lack of suitable graft material and risk of loss of function at the donor site. Additionally, most grafting procedures lead to adhesions, which limit joint mobility. Mechanical mismatch and poor tissue integration, and possible necrosis after implantation, may also occur. Allografts are also alternative therapies for tendon and ligament repair, although they carry the risk of disease transmission and tissue rejection (Rodrigues et al., 2013).

Tissue-engineered tendons (TETs) have been developed as a promising alternative to traditional strategies. They must be able to tolerate the large forces experienced during activities of daily living and exercise. TETs must also be strong enough to bear a load and to avoid deformation and rupture (Shearn et al., 2011).

### 8.5.1 Cells in TETs

Cell sources for TETs include autologous cells such as tenocytes. However, tenocytes are scarce cells, and their isolation causes morbidity at the donor site. Stem cells have huge potential for TETs. ADSCs are good candidates for tendon tissue engineering and can change their phenotype toward a tenogenic direction with appropriate culture conditions (Kraus et al., 2013). However, as these cells are multipotential, there is a risk that they could induce ectopic bone and cartilage formation at the tendon/ligament site. Alternative cell sources include dermal fibroblasts and skeletal muscle-derived cells (MDCs). Dermal fibroblasts have similar characteristics to tenocytes, and can be harvested from a simple skin biopsy (Liu et al., 2006). MDCs are a mixed cell population containing myocytes, fibroblasts, satellite cells, and muscle-derived MSCs, and appear to be suitable cells for development of engineered tendons (Chen et al., 2016).

## 8.5.2 Growth Factors

Several growth factors have been shown to participate in tendon/ligament formation, such as TGF-$\beta$, IGF-1, PDGF, bFGF, or VEGF. When released at the site of injury, these growth factors could play a key role in stimulating local tenocytes (Hsu and Chang, 2004).

## 8.5.3 Scaffolds

A number of natural scaffolds such as silk, collagen, and fibrin can be used to regenerate tendons and ligaments (Yang et al., 2013). However, these scaffolds are often not strong enough to transmit muscle forces. Tissues such as small intestine mucosa, pericardium, or dermis can be processed to remove noncollagen components (which can cause rejection), while retaining the natural collagen structure (predominantly collagen I fibers). However, one limitation of these scaffolds is that their mechanical properties are still significantly lower than those of normal tendons and ligaments (Longo et al., 2010). Decellularized allograft tissues (subsequently recellularized in vitro) are another type of scaffold used in tendon and ligament regeneration. These scaffolds have advantages over allografts: they preserve the natural structure and stability, but also have reduced immunogenicity (Schulze-Tanzil et al., 2012).

Synthetic scaffolds have much stronger mechanical properties than biological scaffolds, but they have limited biocompatibility. Scaffolds can be tailored for use in tissue engineering of tendons. Biodegradable polymers like PGA, PLA, and their copolymer have been proposed for TETs (Reverchon et al., 2012).

## 8.6 CARTILAGE TISSUE ENGINEERING

Cartilage tissue engineering (CTE) is a dynamically changing field that has the potential to address the issues experienced in craniofacial reconstruction. Cartilage lacks an intrinsic regenerative capacity, which makes cartilage reconstruction challenging. Current tissue engineering strategies involve performing a small biopsy to obtain a cartilage specimen, and subsequent isolation of the chondrocytes (cells that maintain the cartilage matrix) through enzymatic digestion of the ECM. The chondrocytes can then be expanded in a two-dimensional monolayer culture to generate large amounts of cartilage. A monolayer culture tends to cause chondrocytes to dedifferentiate or to lose their chondrocytic phenotype. If, however, expanded cells are transferred to a three-dimensional scaffold during that time frame, redifferentiation tends to occur (Chung and Burdick, 2008).

### 8.6.1 Cell Sources for CTE

An important aspect of a tissue-engineered cartilage construct is the cellular component. The use of autologous cells is essential to avoid a potential immunological rejection caused by an allogenic donor cell source. In humans, nasal septum cartilage appears to be a superior source of chondrocytes over articular or auricular cartilage due to a technically easier, more minimally invasive technique of harvesting the cartilage. They are also thought to display increased cartilage formation and better mechanical stability in the resulting neocartilage (Kafienah et al., 2002; Naumann et al., 2004). However, although autologous chondrocytes have been utilized in much of the CTE literature, they have a finite supply; therefore, there has been interest in the potential of stem cells as an alternative cell source. MSCs isolated from bone marrow or adipose tissue are an important source of cells for CTE due to their easy access and high capacity of in vitro expansion (Vinatier et al., 2009). Human adipose tissue could potentially provide stem cell isolates for CTE (Wu et al., 2013). Mahmoudifar and Doran used human adult ADSCs under dynamic culture in PGA scaffolds with combinations of growth factors (TGF-β1, BMP-6, and FGF-2) to induce cartilage synthesis (Mahmoudifar and Doram, 2013). Guasti et al. used human ADSCs in combination with polyhedral oligomeric silsesquioxane poly(carbonate-urea) urethane (POSS-PCU) nanoscaffolds to generate cartilage tissue (Guasti et al., 2014). It is anticipated that with advances in our understanding of chondrogenesis and regulation of stem cell use in cartilage, stem cells can replace chondrocytes as a far superior cell source.

## 8.6.2 Signaling
### 8.6.2.1 Growth Factors
Various growth factors may need to be added to the culture medium to enhance chondrocytic proliferation. IGF-1, FGF, PDGF, the TGF-β family, epidermal growth factor, and members of the BMP family have been shown to enhance chondrocyte proliferation (Blunk et al., 2002).

### 8.6.2.2 Mechanical Stimuli
Mechanical stimuli, such as shear stress, have been shown to increase cartilage matrix expression. Direct compressive loading has also been shown to mechanically stimulate the cells. Shear or direct compression can cause membrane deformation and activate mechanosensitive ion channels. This causes changes in intracellular ionic concentrations, which can activate or suppress various genetic responses (Zhang et al., 2009).

## 8.6.3 Scaffolds
Cartilage reconstruction requires a specific scaffold design. The structural architecture of the scaffold should mimic the exact shape of the native tissue and should support the attachment, proliferation, and differentiation of the desired cell type. The scaffold must also be strong enough to stabilize the reconstructed cartilage until the newly synthesized ECM attains full mechanical stability and function. The rigidity of the scaffold is important from a functional point of view, particularly in structures that enable nasal breathing such as the septum, which is responsible for the shape and tension of its surrounding structures (eg, alar cartilages, columella) (Chung and Burdick, 2008).

To date, a wide range of natural and synthetic materials have been investigated as scaffolding for cartilage repair. Naturally occurring molecules include fibrin, agarose, alginate, chitosan, and collagen. Synthetic polymers include PLA, PGA, and their copolymer, PLGA. To achieve better biocompatibility, researchers are developing nanostructured scaffolds with encapsulated cells (eg, chondrocytes and progenitor cells). Several scaffold-free techniques for generating cartilage have also been investigated (Zhang et al., 2009).

Recent advances in nanotechnology have resulted in the emergence of advanced novel materials with improved properties capable of being used in several biomedical applications, such as the nanocomposite material POSS-PCU (Kannan et al., 2005).

## 8.7 SUMMARY

Bone is a dynamic structure, and any disease or injury affecting it can have life-changing consequences for the patient. The incidence of bone disease is increasing due to an increase in the aging population. BTE is a much-needed alternative to conventional bone grafts. BTE is a complex, dynamic process that begins with recruitment, proliferation, and differentiation of osteoprogenitor cells. Advances in tissue engineering are achieved by creating optimal porous scaffolds that provide mechanical support. The future of BTE lies in the creation of a cost-effective, mechanically strong porous construct with the ability to effectively integrate into the host while retaining adequate vascularization.

## REFERENCES

Ahmed, E.A., 2015. Hydrogel: preparation, characterization, and applications: a review. J. Adv. Res. 6 (2), 105–121.

Alvarez, K., Nakajima, H., 2009. Metallic scaffolds for bone regeneration. Materials 2 (3), 790–832.

Amini, A.R., Laurencin, C.T., Nukavarapu, S.P., 2012. Bone tissue engineering: recent advances and challenges. Crit. Rev. Biomed. Eng. 40 (5), 363–408.

Aubin, J.E., 2001. Regulation of osteoblast formation and function. Rev. Endocr. Metab. Disord. 2 (1), 81–94.

Bielby, R.C., Boccaccini, A.R., Polak, J.M., Buttery, L.D.K., 2004. In vitro differentiation and in vivo mineralization of osteogenic cells derived from human embryonic stem cells. Tissue Eng. 10 (9-10), 1518–1525.

Bilousova, G., Jun, D.H., King, K.B., Langhe, S.D., Chick, W.S., Torchia, E.C., et al., 2011. Osteoblasts derived from Induced Pluripotent Stem Cells form Calcified Structures in Scaffolds both in vitro and in vivo. Stem Cells 29 (2), 206–216.

Birmingham, E., Niebur, G.L., McHugh, P.E., Shaw, G., Barry, F.P., McNamara, L.M., 2012. Osteogenic differentiation of mesenchymal stem cells is regulated by osteocyte and osteoblast cells in a simplified bone niche. Eur. Cell. Mater. 23, 13–27.

Blunk, T., Sieminski, A.L., Gooch, K.J., Courter, D.L., Hollander, A.P., Nahir, A.M., et al., 2002. Differential effects of growth factors on tissue-engineered cartilage. Tissue Eng. 8 (1), 73–84.

Bose, S., Roy, M., Bandyopadhyay, A., 2012. Recent advances in bone tissue engineering scaffolds. Trends Biotechnol. 30 (10), 546–554.

Boyle, W.J., Simonet, W.S., Lacey, D.L., 2003. Osteoclast differentiation and activation. Nature 423 (6937), 337–342.

Caplan, A.I., 1991. Mesenchymal stem cells. J. Orthop. Res. 9 (5), 641–650.

Chen, B., Ding, J., Zhang, W., Zhou, G., Cao, Y., Liu, W., et al., 2016. Tissue engineering of tendons: a comparison of muscle-derived cells, tenocytes, and dermal fibroblasts as cell sources. Plas. Reconstr. Surg. 137 (3), 536e–544ee.

Chung, C., Burdick, J.A., 2008. Engineering cartilage tissue. Adv. Drug. Deliv. Rev. 60 (2), 243–262.

Clarke, B., 2008. Normal bone anatomy and physiology. Clin. J. Am. Soc. Nephrol. 3 (Suppl. 3), S131–S139.

Cowan, C.M., Shi, Y.Y., Aalami, O.O., Chou, Y.F., Mari, C., Thomas, R., et al., 2004. Adipose-derived adult stromal cells heal critical-size mouse calvarial defects. Nat. Biotechnol. 22 (5), 560–567.

Dai, J., Rabie, A.B., 2007. VEGF: an essential mediator of both angiogenesis and endochondral ossification. J. Dent. Res. 86 (10), 937–950.

Das, K., Balla, V.K., Bandyopadhyay, Bose, S., 2008. Surface modification of laser-processed porous titanium for load-bearing implants. Scripta Materialia 59 (8), 822–825.

Dhandayuthapani, B., Yoshida, Y., Maekawa, Kumar, D.S., 2011. Polymeric scaffolds in tissue engineering application: a review. Int. J. Polym. Sci.1687–9422.

Eriksen, E.F., Axelrod, D.W., Melsen, F., 1994. Bone Histomorphometry. Raven Press, New York, NY, pp. 1–12.

Fu, Q., Saiz, E., Rahaman, M., Tomsia, A., 2011. Bioactive glass scaffolds for bone tissue engineering: state of the art and future perspectives. Mater. Sci. Eng. 31 (7), 1245–1256.

Gilbert, S.F., Sunderland, M.A., 2000. Developmental Biology., sixth ed. Sinauer Associates, Sunderland, MA.

Govoni, K.E., Wergedal, J.E., Florin, L., Angel, P., Baylink, D.J., Mohan, S., 2007. Conditional deletion of insulin-like growth factor-I in collagen type 1alpha2-expressing cells results in postnatal lethality and a dramatic reduction in bone accretion. Endocrinology 148 (12), 5706–5715.

Grottkau, B.E., Lin, Y., 2013. Osteogenesis of adipose-derived stem cells. Bone Res. 1 (2), 133–145.

Guasti, L., Vagaska, B., Bulstrode, N.W., Seifalian, A.M., Ferretti, P., 2014. Chondrogenic differentiation of adipose tissue-derived stem cells within nanocaged POSS-PCU scaffolds: a new tool for nanomedicine. Nanomedicine 10 (2), 279–289.

Hoemann, C.D., El-Gabalawy, H., McKee, M.D., 2009. In vitro osteogenesis assays: influence of the primary cell source on alkaline phosphatase activity and mineralization. Pathol. Biol. 57 (4), 318–323.

Hollinger, J.O., Hart, C.E., Hirsch, S.N., Lynch, S., Friedlaender, G.E., 2008. Recombinant human platelet-derived growth factor: biology and clinical applications. J. Bone Joint. Surg. Am. 90 (Suppl. 1), 48–54.

Hsu, C., Chang, J., 2004. Clinical implications of growth factors in flexor tendon wound healing. J. Hand Surg. Am. 29 (4), 551–563.

Huang, Z., Nelson, E.R., Smith, R.L., Goodman, S.B., 2007. The sequential expression profiles of growth factors from osteoprogenitors [correction of osteroprogenitors] to osteoblasts in vitro. Tissue Eng. 13 (9), 2311–2320.

Hwang, N.S., Zhang, C., Hwang, Y.S., Varghese, S., 2009. Mesenchymal stem cell differentiation and roles in regenerative medicine. Wiley Interdiscip. Rev. Svst. Bio. Med. 1 (1), 97–106.

Jaiswal, N., Haynesworth, S.E., Caplan, A.I., Bruder, S.P., 1997. Osteogenic differentiation of purified, culture-expanded human mesenchymal stem cells in vitro. J. Cell. Biochem. 64 (2), 295–312.

Kafienah, W., Jakob, M., Demarteau, O., Frazer, A., Barker, M.D., Martin, I., et al., 2002. Three-dimensional tissue engineering of hyaline cartilage: comparison of adult nasal and articular chondrocytes. Tissue Eng. 8 (5), 817–826.

Kahn, A.J., Simmons, D.J., 1975. Investigation of cell lineage in bone using a chimaera of chick and quail embryonic tissue. Nature 258, 325–327.

Kannan, R.Y., Salacinski, H.J., Butler, P.E., Seifalian, A.M., 2005. Polyhedral oligomeric silsesquioxane nanocomposites: the next generation material for biomedical applications. Acc. Chem. Res. 38 (11), 879–884.

Kasten, P., Beyen, I., Niemeyer, P., Luginbuhl, R., Bohner, M., Richter, W., 2008. Porosity and pore size of beta-tricalcium phosphate scaffold can influence protein production and osteogenic differentiation of human mesenchymal stem cells: an in vitro and in vivo study. Acta Biomater. 4 (6), 1904–1915.

Kawaguchi, H., Oka, H., Jingushi, S., Izumi, T., Fukunaga, M., Sato, K., et al., 2010. A local application of recombinant human fibroblast growth factor 2 for tibial shaft fractures: a randomized, placebo-controlled trial. J. Bone Miner. Res. 25 (12), 2735–2743.

Kraus, A., Woon, C., Raghavan, S., Megerle, K., Pham, H., Chang, J., 2013. Co-culture of human adipose-derived stem cells with tenocytes increases proliferation and induces differentiation into a tenogenic lineage. Plast. Reconstr. Surg. 132 (5), 754e–766e.

Lee, T.J., Jang, J., Kang, S., Jin, M., Shin, H., Kim, D.W., et al., 2013. Enhancement of osteogenic and chondrogenic differentiation of human embryonic stem cells by mesodermal lineage induction with BMP-4 and FGF2 treatment. Biochem. Biophys. Res. Commun. 430 (2), 793–797.

Liu, W., Chen, B., Deng, D., Xu, F., Cui, L., Cao, Y., 2006. Repair of tendon defect with dermal fibroblast engineered tendon in a porcine model. Tissue Eng. 12 (4), 775–788.

Longo, U.G., Lamberti, A., Maffulli, N., Denaro, V., 2010. Tendon augmentation grafts: a systematic review. Br. Med. Bull. 94 (1), 165–188.

Mackay, A.M., Beck, S.C., Murphy, J.M., Barry, F.P., Chichester, C.O., Pittenger, M.F., 1998. Chondrogenic differentiation of cultured human mesenchymal stem cells from marrow. Tissue Eng. 4 (4), 415–428.

Mahmoudifar, N., Doran, P.M., 2013. Osteogenic differentiation and osteochondral tissue engineering using human adipose-derived stem cells. Biotech. Prog. 29 (1), 176–185.

Makhdom, A.M., Hamdy, R.C., 2012. The role of growth factors on acceleration of bone regeneration during distraction osteogenesis. Tissue Eng. Part B. Rev. 19 (5), 442–453.

Mastrogiacomo, M., Scaglione, S., Martinetti, R., Dolcini, L., Beltrame, F., Cancedda, R., et al., 2006. Role of scaffold internal structure on in vivo bone formation in macroporous calcium phosphate bioceramics. Biomaterials 27 (17), 3230–3237.

Meyer, U., Meyer, Th, Handschel, J., Weismann, H.P., 2009. Fundamentals of Tissue Engineering and Regenerative Medicine. Springer-Verlag, Berlin.

Miyazono, K., Kamiya, Y., Morikawa, M., 2010. Bone morphogenetic protein receptors and signal transduction. J. Biochem. 147 (1), 35–51.

Mundy, G.R., Martin, T.J., 1993. Physiology and Pharmacology of Bone. Handbook of Experimental Physiology. Springer-Verlag, Berlin.

Nakamura, K., Kurokawa, T., Kato, T., Okazaki, H., Ma-mada, K., Hanada, K., et al., 1996. Local application of basic fibroblast growth factor into the bone increases bone mass at the applied site in rabbits. Arch. Orthop. Trauma. Surg. 115 (6), 344–346.

Naumann, A., Dennis, J.E., Aigner, J., Coticchia, J., Arnold, J., Berghaus, A., 2004. Tissue engineering of autologous cartilage grafts in three-dimensional in vitro macroaggregate culture system. Tissue Eng. 10 (11–12), 1695–1706.

Netter, F.H., 1987. Musculoskeletal system: anatomy, physiology, and metabolic disorders. Ciba-Geigy Corporation, Summit, New Jersey.

Nguyen, L.H., Annabi, N., Nikkhah, M., Bae, H., Binan, L., Park, S., et al., 2012. Vascularized bone tissue engineering: approaches for potential improvement. Tissue Eng. Part B. Rev. 18 (5), 363–382.

Ochman, S., Frey, S., Raschke, M.J., Deventer, J.N., Meffert, R.H., 2011. Local application of VEGF compensates callus deficiency after acute soft tissue trauma—results using a limb-shortening distraction procedure in rabbit tibia. J. Orthop. Res. 29 (7), 1093–1098.

Pagkalos, J., Cha, J.M., Kang, Y., Heliotis, M., Tsiridis, E., Mantalaris, A., 2010. Simvastatin induces osteogenic differentiation of murine embryonic stem cells. J. Bone Miner. Res. 25 (11), 2470–2478.

Park, J.B., 2011. The use of hydrogels in bone-tissue engineering. Med. Oral. Patol. Oral. Cir. Bucal. 16 (1), e115–e118.

Pittenger, M.F., Mackay, A.M., Beck, S.C., Jaiswal, R.K., Douglas, R., Mosca, J.D., et al., 1999. Multilineage potential of adult human mesenchymal stem cells. Science 284 (5411), 143–147.

Quarles, L.D., Yohay, D.A., Lever, L.W., Caton, R., Wenstrup, R.J., 1992. Distinct proliferative and differentiated stages of murine MC3T3-E1 cells in culture: an in vitro model of osteoblast development. J. Bone Miner. Res. 7 (6), 683–692.

Reverchon, E., Baldino, L., Cardea, S., De Marco, I., 2012. Biodegradable synthetic scaffolds for tendon regeneration. Muscles Ligaments Tendons J. 2 (3), 181–186.

Rodrigues, M.T., Reis, R.L., Gomes, M.E., 2013. Engineering tendon and ligament tissues: present developments towards successful clinical products. J. Tissue Eng. Regen. Med. 7 (9), 673–686.

Roosa, S.M., Kemppainen, J.M., Moffitt, E.N., Krebsbach, P.H., Hollister, S.J., 2010. The pore size of polycaprolactone scaffolds has limited influence on bone regeneration in an in vivo model. J. Biomed. Mater. Res. A. 92 (1), 359–368.

Schulze-Tanzil, G., Al-Sadi, O., Ertel, W., Lohan, A., 2012. Decellularized tendon extracellular matrix—a valuable approach for tendon reconstruction? Cells 1 (4), 1010–1028.

Seifert, R.A., Hart, C.E., Phillips, P.E., Forstrom, J.W., Ross, R., Murray, M.J., et al., 1989. Two different sub-units associate to create isoform-specific platelet-derived growth factor receptors. J. Biol. Chem. 264 (15), 8771–8778.

Shearn, J.T., Kinneberg, K.R.C., Dyment, N.A., Gallowat, M.T., Kenter, K., Wylie, C., et al., 2011. Tendon tissue engineering: progress, challenges, and translation to the clinic. J. Musculoskelet. Neuronal. Interact. 11 (2), 163–173.

Shimko, D.A., Burks, C.A., Dee, K.C., Nauman, E.A., 2004. Comparison of in vitro mineralization by murine embryonic and adult stem cells cultured in an osteogenic medium. Tissue Eng. 10 (9–10), 1386–1398.

Sottile, V., Thomson, A., McWhir, J., 2003. In vitro osteogenic differentiation of human ES cells. Cloning Stem Cells 5 (2), 149–155.

Tashiro, K., Inamura, M., Kawabata, K., Sakurai, F., Yamanishi, K., Hayakawa, T., et al., 2009. Efficient adipocyte and osteoblast differentiation from mouse induced pluripotent stem cells by adenoviral transduction. Stem Cells 27 (8), 1802–1811.

Vinatier, C., Bouffi, C., Merceron, C., et al., 2009. Cartilage tissue engineering: towards a biomaterial-assisted mesenchymal stem cell therapy. Curr. Stem Cell Res. Ther. 4 (4), 318–329.

Vo, T.N., Kasper, K., Mikos, A., 2012. Strategies for controlled delivery of growth factors and cells for bone regeneration. Adv. Drug. Deliv. Rev. 64 (12), 1292–1309.

Wang, P., Zhao, L., Liu, J., Weir, M.D., Zhou, X., Xu, H.H.K., 2014. Bone tissue engineering via nanostructured calcium phosphate biomaterials and stem cells. Bone Res. 2, 14017.

Woll, N.L., Bronson, S.K., 2006. Analysis of embryonic stem cell-derived osteogenic cultures. Methods Mol. Biol. 330, 149–159.

Wu, C., Zhou, Y., Fan, W., Han, P., Chang, J., Yuen, J., et al., 2012. Hypoxia-mimicking mesoporous bioactive glass scaffolds with controllable cobalt ion release for bone tissue engineering. Biomaterials 33, 2076–2085.

Wu, L., Cai, X., Zhang, S., Karperien, M., Lin, Y., 2013. Regeneration of articular cartilage by adipose tissue derived mesenchymal stem cells: perspectives from stem cell biology and molecular medicine. J. Cell. Physiol. 228 (5), 938–944.

Yang, G., Rothrauff, B.B., Tuan, R.S., 2013. Tendon and ligament regeneration and repair: clinical relevance and developmental paradigm. Birth Defects Res. C Embryo Today 99 (3), 203–222.

Zhang, L., Hu, J., Athanasiou, K.A., 2009. The role of tissue engineering in articular cartilage repair and regeneration. Crit. Rev. Biomed. Eng. 37 (1–2), 1–57.

Zhang, M., Xuan, S., Bouxsein, M.L., von Stechow, D., Akeno, N., Faugere, M.C., et al., 2002. Osteoblast-specific knockout of the insulin-like growth factor (IGF) receptor gene reveals an essential role of IGF signaling in bone matrix mineralization. J. Biol. Chem. 277 (46), 44005–44012.

# INDEX

*Note*: Page numbers followed by "*b*," "*f*," and "*t*" refer to boxes, figures, and tables, respectively.

Sensory corpuscles, 25
Shear stress, 62, 92
Skin grafts, 17, 24
Solvent casting/particulate leaching (SCPL), 7*t*
Sphincter, 3–4
Spinal cord, 29–33
Split skin grafts, 17–20, 20*f*
Synthetic scaffolds, 90

**T**
Tendon tissue engineering, 89–90
Tenocytes, 10, 89–90
Tissue engineering, 1, 3
Tissues, 1–2, 6, 90
Trabecular bone, 78
Trachea, 1–2
Transforming growth factor beta (TGF-β), 10, 83, 86*t*

**V**
Valvular tissue engineering, 61
Vascular endothelial growth factors (VEGF), 10, 24, 60, 80, 86*t*, 89
Vascularization, 11–13, 24, 61, 89
Vitreous humor, 44*f*, 45

**W**
Wallerian degeneration, 36–37
Wounds, 18–20, 19*t*

**X**
Xenogenic cells, 8

**Y**
Young's modulus, 51, 87, 88*b*

Printed in the United States
By Bookmasters